APPLICATIONS OI BLOCKCHAIN TECHNOLOGY
An Industry Focus

Anita Ravani

Graduate School of Education and Psychology
Pepperdine University, Los Angeles, CA, USA

Sashi Edupuganti

School of Business
Pepperdine University, Malibu, CA, USA

Jeannette Pugh

Graduate School of Education and Psychology
Pepperdine University, Los Angeles, CA, USA

Sooraj Sushama

Graduate School of Education and Psychology
Pepperdine University, Los Angeles, CA, USA

CRC Press
Taylor & Francis Group
Boca Raton London New York

CRC Press is an imprint of the
Taylor & Francis Group, an **informa** business

First edition published 2024
by CRC Press
2385 NW Executive Center Drive, Suite 320, Boca Raton FL 33431

and by CRC Press
4 Park Square, Milton Park, Abingdon, Oxon, OX14 4RN

CRC Press is an imprint of Taylor & Francis Group, LLC

Library of Congress Cataloging-in-Publication Data (applied for)

ISBN: 978-1-032-47088-7 (hbk)
ISBN: 978-1-032-47157-0 (pbk)
ISBN: 978-1-003-38480-9 (ebk)

DOI: 10.1201/9781003384809

Typeset in Times New Roman
by Innovative Processors

This volume is humbly dedicated to industry leaders and students who are seeking answers to revolutionary technology innovations to transform their understanding of blockchain technology.

Preface

The authors have prudently crafted this book to effectively disseminate blockchain technology and its application for the benefit of all readers.

This book unpacks the confusion around blockchain technology and offers the readers a carefully crafted strategy for the real-world application of blockchain technology.

Highlights of this book include the following:

- **Deconstruction of Blockchain Technology:** Deconstruction of complex technology to provide deeper insight into its design and framework.
- **Application of Blockchain Technology:** Detailed framework showing the practical application of blockchain technology in different industries.
- **Rich Practical Examples:** Several samples of industry examples utilizing blockchain technology.
- **Case Studies:** Inclusion of several case study analyses to provide a better understanding of the practical application of blockchain technology.

Anita Ravani
Sashi Edupuganti
Jeannette Pugh
Sooraj Sushama

List of Figures

Contents

Part II: Implementation, Advantages and Disadvantages of Blockchain

Introduction

Every individual, organization, nation, and idea will follow a bell curve, those who choose to recognize the peak and start a new journey that will be relevant in the future.

Human society has evolved. From hunter-gatherer societies to agrarian societies, humans have learned to cooperate and work together to produce food and goods. The development of agriculture and the domestication of animals allowed for the growth of settled communities and the creation of a surplus of resources which in turn led to the development of social hierarchies, including a feudal system and more democratic forms of government. Throughout this transformational journey, humans have also developed new technologies and systems of communication, which have further enabled cooperation and the exchange of ideas and goods.

The Industrial Revolution and the subsequent Information Age have brought about significant changes in society, including new modes of transportation, communication, and production that enabled humans to achieve levels of productivity and efficiency previously unimaginable. The resulting increase in wealth and prosperity, combined with improvements in healthcare and living conditions, contributed to a population explosion that transformed society. As we continue to evolve as a society, it is important to recognize and build upon the lessons of our past, while also adapting to new challenges and opportunities. This requires ongoing cooperation, planning, and innovation.

After multiple iterations of refining the agrarian world from a period of stagflation, we created the industrial revolution that brought together the ingenuity of the human mind in producing leverage as we have never seen before. The ability to travel faster, do it on a larger scale than the size of our biological limitations, and more importantly, produced a mechanism to protect ourselves which started the population explosion during this stage. This evolution brought together the focus on process as a commodity to build scale. However, this wasn't all good as it went through the exploitation of humans, creating the need for a more balanced economic model by creating unions to balance the world.

While the pull and push of the industrial revolution went through its cycles with world wars, economic meltdowns, and reorganization of the world order after a brief period of stability, we are currently in the world of a digital revolution where we are connected digitally with one another without ever getting to meet each other. This evolution has created information as a commodity for exploitation. This has peaked as in previous revolutions with an imbalanced ecosystem with 20 companies managing 80% of the digital footprint which has created a significant impact on basic human rights, including digital rights. The digital revolution has transformed how we communicate, work, and live. The need for a new revolution is upon us and is being called the social revolution. Social revolution is the mobilization of people to demand greater equality, social justice, and environmental sustainability. This movement is fueled by a growing awareness of the negative impacts of the current economic and political system, as well as the opportunities presented by new technologies and social media to organize and communicate.

One of the key challenges of this social revolution is to ensure that it is inclusive and representative of diverse voices and perspectives. This requires not only the empowerment of marginalized communities but also the development of new forms of governance and decision-making that is more democratic and participatory. Social revolution is about creating a more just and equitable society that values human rights, diversity, and sustainability. It will require the engagement and participation of people from all walks of life and the development of new technologies, policies, and institutions that support these values.

The digital transformation of society is a shift towards a more digital, data-driven world in which every organization is a digital organization and every individual is a digital citizen. The transformation has been driven by the ubiquity of technology and the adoption of mobile devise and applications. However, it has also led to the exploitation of data and the need for a more conscious ecosystem. The social revolution and the digital transformation of society are intertwined because the social revolution seeks to address the imbalances and issues created by digital transformation. It aims to create a more equitable and consensus-based ecosystem that protects digital rights and promotes the collective good. Blockchain technology's adoption and evolution are key to achieving this goal.

The social revolution is where the balance between the digital have-nots and digital haves is balanced through a more consensus and context-based ecosystem that was accidentally but timely triggered by the technology behind cryptocurrency called the blockchain. We are at the start of a journey for the next few decades. We can ensure that the adoption, evolution, and application of this technology is for the collective good and benefit of the new world order. In this section, four main points are covered:

1. Every organization is a digital organization.
2. Every individual is a digital citizen.

3. Digital rights are human rights.
4. Digital transformation impact on organizations and individuals

Every Organization is a Digital Organization

Three decades ago, the mantra was "Every organization is a technology organization", a decade ago the mantra was "Every organization is a digital organization", now the new mantra is "Every organization is a data/conscience organization". Why do we say that? From Fortune 100 companies to a foot cart in Manhattan, everybody uses technology, specifically digital technology to do their business, acquire consumers and build a relationship with them, be it a one-person entity or a million-person entity; they have a digital strategy and footprint that produces loads and loads of data. These days every organization has become a data-driven organization and the misuse of data has created the need for more accountability in the collection and use of that data requiring a more conscious ecosystem that creates the need for 3rd party entities to act as a consensus broker rather than the 2-party transactional system that exists today. This is the design of the blockchain.

Every Individual is a Digital Citizen

Today people do not have landlines, and everybody is comfortable with a cell phone which has its advantages and disadvantages. Cell phone ubiquity and its adoption have promoted the creation of a very healthy application of the ecosystem that ranges from health monitoring to food ordering, and even to managing finances. Today there are roughly around 10 million mobile applications that embrace this new technology, especially in the evolving world there has been a rapid rise due to the lack of legacy anchors. This means that every individual is a digital citizen, and technology is another layer added to the history of our evolution which is now a permanent part of our fabric and is here to stay.

Digital Rights are Human Rights

The characteristics of human rights are universal and inalienable, indivisible, independent, equal, and nondiscriminatory, and this holds for both rights and obligations. Based on these criteria, the universally accepted rights are rights to life and liberty, freedom from slavery and torture, freedom of opinion and expression, and the right to work and education. Grounded in the guiding principles of what has been categorized as human rights, there is an evolving acceptance of digital rights as human rights. Accepting digital rights and human rights is important both as an individual and as an enterprise. Centered on the digital footprint there is a whole lot that can be inferred about an individual and that could be used for good or bad, today it is used one-sided by large enterprises.

Digital Transformation Impact on Organizations and Individuals

As digital becomes the new path that transcends the world of connectivity, to establish balance, governments across the world are introducing and quickly evolving a new set of regulations around privacy and security and the framework of how and what can be done to ensure them. Hefty fines that could bring an organization to its knees when there is a breach or a compliance violation are meant to be the stick. The carrot is Web 3.0 which is enabled by blockchain for organizations to use the opportunity of compliance to start a new evolution of the digital footprint that can build a new ecosystem that makes investment capital rather than operational. This is important for many reasons, including taxation among other things. For individuals, it means they get to have better control of their digital footprint without being at the mercy of the legal language or the noise that is currently created. As individuals, we are destined to lose our critical thinking due to the asymmetric nature of information analysis available today.

Trust ability is the foundation for a future digital world. Every business is in the business of "Trust". The cost of acquiring and retaining a consumer is high, the value of acquiring a consumer is beyond the first transaction and becomes valuable only when there is repeat business over a long period. The cost of retention becomes high when trust becomes a question, as can be noticed from the Cambridge analytical case, the trust in Facebook as a platform went down the drain and hasn't been recouped.

Formula: Immutability + Consensus = New Digital Identity

This brings us to the formula required to embark on a new world of social evolution, overcoming the inability to trust the information provided to us that is widely validated across non-transacting parties, and that is exactly what blockchain does for you.

Background: When the World Wide Web specification was created and released in the 1990s, its vision was to create a more collaborative, transparent, and self-evolving network. The vision was however embraced based on the maturity of the market as it existed at that time and has since evolved in a direction that is against the original vision that it was created for. This misdirection has been recognized by none other than the person responsible for the WWW specification, Sir Tim Berners-Lee, and he is currently working on a new network that realigns with the original vision. Irrespective of this, the specification provided nonmaterial addition to global commerce as mentioned below.

- **Web 1.0:** For those of you who remember, the first version of the internet was websites and blogs, which were used mostly as a medium for content distribution. This evolution, even with its nascent changes has introduced 1.9T in net growth.

- **Web 2.0:** Which is what we primarily are on today is a commerce web with the likes of Amazon, Alibaba, and Facebook where it was bidirectional but primarily built to build a commerce engine with information being a new commodity for trading. This evolution by its own measure has added about 5.6T to the world economy.
- **Web 3.0:** As was mentioned at the start of the chapter in the arch model every journey peak before starting to diminish, such is the case for web 2.0 where the disadvantages of it have surpassed its advantages of it and are ready for a new cycle of evolution termed the Web 3.0—the decentralized web. While we are still in the early stages of it, the speed at which it is being embraced is sizing up this opportunity to be around 15 to 20 trillion US dollars.

Digital Twin

The concept of consumer360 or customer360 has become increasingly popular among organizations as they strive to better understand and cater to their customers' needs. However, this has also resulted in a power imbalance where organizations have access to a vast amount of data on their customers, while customers themselves may not have access to the same level of information about themselves or their interactions with these organizations. This power imbalance can have a significant impact on areas such as media and children, where unreliable or inaccurate information can be spread quickly and easily. To address this, there is a need for individuals to have greater control over their data and the ability to share it based on their consent. As we move towards Web 3.0, individuals are gaining more rights to the content they create and could choose how it is used or monetized. This shift towards greater individual control over data and content is a positive development that can help to rebalance power dynamics and ensure that individuals have greater agency and autonomy in their interactions with organizations.

Part I, comprising of 5 chapters, focuses on *Blockchain Frameworks*

- In Chapter 1, Ravani aims to provide insights into the construction and core competencies of blockchain technology. The technology age is perpetually evolving. This chapter offers an engaging view of blockchain technology and describes blockchain technology from a technology viewpoint. The focus of the chapter centers on blockchain architecture, the benefits of blockchain, the important features of blockchain, and understanding the basic technology surrounding blockchain. This chapter conducts a comprehensive analysis of introducing relevant content of blockchain technology, provides fundamental principles of blockchain core structure and elements, and underscores opportunities of blockchain technology. The chapter is organized into seven sections: introduction, structure, application, conclusion, chapter summary, takeaway lessons, and chapter exercises.
- In Chapter 2, Ravani explores the possibilities of blockchain technology in the finance industry. The chapter dives into the importance of the regulatory

framework surrounding the current industry and the lack of regulatory insight around cryptocurrency that has caused a steady increase in cybercriminal activities. The chapter highlights the benefits of blockchain technology in the finance industry and provides a rich framework to dissect the use of the technology. Furthermore, content from rich case studies is explored. Christie's auction house's successful implementation of blockchain during the COVID-19 panacea and the success of non-fungible tokens (NFTs) is discussed. The chapter is organized into eight sections: introduction, structure, framework, application, conclusion, chapter summary, takeaway lessons, and chapter exercises.

- In Chapter 3, Pugh and Sushama, showcase the integration and benefits of blockchain technology in the healthcare industry. The chapter offers an innovative framework for blockchain use in healthcare and directs the reader to the many opportunities of blockchain technology in the healthcare industry. The chapter focuses on improved data security, enhanced patient care, improved interoperability, reduced costs, medical records, prescription drugs, clinical trials, and exhibits the importance of blockchain-based solutions to develop new healthcare-related applications that improve patient care while reducing costs and increasing transparency for the entire healthcare ecosystem. Two relevant case studies are included to facilitate blockchain comprehension and application in the healthcare industry. The chapter is organized into eight sections: introduction, structure, framework, application, conclusion, chapter summary, takeaway lessons, and chapter exercises.
- In Chapter 4, Ravani and Edupuganti aver the use of blockchain technology and its relevance in the law enforcement industry. The chapter highlights the current deficiencies in regulations concerning digital assets and important topics such as money laundering, KYC, and decentralized finance (DeFi) are deliberated. The benefits of blockchain-based solutions are discussed based on two main characteristics of law enforcement accountability and justice. An inclusive framework is provided for the readers to fully comprehend the pragmatic use of blockchain technology in the existing governance framework. Examples of current blockchain implementation in the law enforcement industry are revealed to build upon the reader's understanding of blockchain successes in the industry. The chapter is organized into eight sections: introduction, structure, framework, application, conclusion, chapter summary, takeaway lessons, and chapter exercises.
- In Chapter 5, Ravani opines on the inclusion of blockchain technology in the supply chain management industry. The chapter displays the struggles of traditional supply chain management systems and actively approaches blockchain-based solutions. The use of smart contracts is unveiled to provide the audience with a framework that incorporates all stakeholders in the supply chain. The inclusion of supply chains from various industries such as global car and automobile manufacturing, global fruit and vegetable processing, global pharmaceuticals and medicine manufacturing, global

sugar manufacturing, global cheese manufacturing, and global cosmetics manufacturing are discussed to galvanize supply chain importance. Two case studies are dissected to show a practical application of blockchain technology in the industry. The chapter is organized into eight sections: introduction, structure, framework, application, conclusion, chapter summary, takeaway lessons, and chapter exercises.

Part II, comprising of 5 chapters, focuses on *Implementation, Advantages and Disadvantages of Blockchain*

- In Chapter 6, Ravani and Sushama showcase the structural implementation of blockchain technology. The chapter focuses on the structural architecture of blockchain technology and dissects the consensus algorithms that are used in the technology. Consensus algorithms enable multiple nodes in a network to come to a mutual agreement on the current state of the blockchain without the need for a central authority. There are several consensus algorithms explained in this chapter that are used in blockchain technology, including Proof of Work (PoW), Proof of Stake (PoS), Delegated Proof of Stake (DPoS), and Byzantine Fault Tolerance (BFT). Furthermore, each of these algorithms has its own set of advantages and disadvantages in terms of security, scalability, and energy efficiency that is included in the chapter discussion network. The chapter concludes with highlights of the advantages and disadvantages of each type of consensus mechanism and provides the user/reader with clarity on their benefits.
- In Chapter 7, Ravani delves into blockchain opportunities and vulnerabilities. The chapter focuses on how public ledger technology has been a part of the history of technological evolution. The avantages of early adopters of blockchain technology who have experienced breakthroughs in many daily activities and business processes are discussed. The infrastructure of blockchain is explored to understand the value of an ethically sourced shared repository of data for businesses. Immutability and decentralization which are core functions of blockchain technology are discussed. Moreover, smart contracts are discussed in the chapter to offer the reader a more in depth understanding of the technology and its use. The chapter includes many opportunities that can leapfrog industries using blockchain, but also critically analyzes some disadvantages of blockchain technology and provides the reader with an in depth understanding of where the technology shortfall lies.
- In Chapter 8, Ravani provides insight into blockchain exploitation. The chapter outlines the use of ledgers that have been around for thousands of years to keep track of transactions and builds a case for today's ledgers that have evolved into digital and computerized formats. The chapter includes a thoughtful explanation of ledger technology and the future of ledger technology. Furthermore, the strengths and weaknesses of blockchain technology are explored. The basic components of blockchain technology are discussed alongside some advantages and disadvantages to consider before

implementing blockchain technology. The chapter reiterates the strengths and the advantages of blockchain infrastructure which is futuristic and sustainable for many industries and provides a deeper understanding of the blockchain architecture for any user/reader lacking a fundamental understanding of blockchain technology.

- In Chapter 9, Ravani and Sushama explore the crossroads of merging business and technology. The chapter successfully defends blockchain technology in various industries and provides rich insight into the success of blockchain use in the current business environment. The merging of blockchain technology is spotlighted in the chapter to impart an understanding of the current active use of blockchain technology in today's business and to further the readers' understanding of the value of the technology in the future. Moreover, the chapter builds on the core fundamentals of the technology. In addition, the chapter includes frameworks such as Porter's Five, design for six sigma to introduce the benefits of blockchain technology, and merging previously created quality standards by various industries to utilize blockchain technology and the next generation technology to improve on preexisting quality control and efficiency models.

- In Chapter 10, Ravani and Sushama conclude the book with blockchain application and its future application possibilities. This final chapter of the book centers on big data technologies and their future potential use. The chapter combines highlights the use of the current blockchain in the supply chain and finance in the industry. Furthermore, a comprehensive SWOT analysis is included on the strengths, weaknesses, opportunities, and threats of blockchain technology that provides the reader with quick and easy to understand value creation of blockchain technology. In this final chapter, there is a culmination of supporting blockchain technology to facilitate change in future business decision making and seriously consider the basis of blockchain technology and its immense power in the future of technology.

About the Authors

Anita Ravani is the founder of APR Investment and Management, a leading management and consulting firm based out of Los Angeles, California that specializes in finance and accounting to drive consumer decision-making. She is a member of senior management at a global financial firm with over 20 years of finance and accounting experience working with Fortune 500 companies to solve their most important financial challenges.

Anita is a two-time inductee of the prestigious Beta Gamma Sigma Honor Society and holds a Ph.D. in Global Leadership and Change from the Graduate School of Education and Psychology at Pepperdine University. Anita is a purposeful and tenaciously dedicated, result-oriented leader and project management professional with deep knowledge and experience in developing and implementing financial systems, processes, and controls that significantly improved P&L scenarios related to multi-faceted markets in telecommunications, manufacturing, aerospace, health care, real estate, and financial services.

Sashi Edugapanti is the founder and CEO of ODE Holdings. A shaper and a creator who enjoys building, transforming, and growing new organizations, solutions, and markets.

Blessed to have the dual experience as a Fortune 100 general manager and a serial entrepreneur. He believes in not only being good at what you do but by being GOOD to others, learning from the past, planning for the future while living in the PRESENT, and being thoughtful about the FUTURE!

Jeannette Pugh is a SHRM-certified, global human resources professional with multi-industry HR generalist, specialist, and line management experience, She collaborates with teams to develop sustainable solutions for organizations.

Jeannette has a proven track record in empowering employee networks, leading transformational retention initiatives, and serving as a strategic business partner within high-performance organizations. She has partnered with leaders to improve morale, reduce turnover, increase positive employee contributions, and lead a significant turnaround in multi-industry, domestic, and international organizations. Jeannette catalyzes for people and organizations to tap into their limitless potential and flourish. Specialties include:

- Recruitment/Retention/Career Development
- Global Leadership Development
- Organizational Development/Assessment
- Employee Relations/Conflict Resolution
- Strategic Succession Planning
- R&D, Sales, and Manufacturing Experience
- Principle-centered leader of high-performing teams
- Employee Engagement
- Diversity, Equity, and Inclusion
- Merger & Acquisition Experience

Sooraj Sushama is an accomplished Executive Advisor for Data Analytics and Data Science, currently leading the charge at a top US health insurance firm. Sooraj has over 18 years of consulting and management experience in Data Analysis and Business Strategy, specifically focusing on the Health Insurance, Education, and Finance industries. Sooraj's work involves providing strategic guidance to executive leadership to make data-driven business decisions.

Sooraj's educational background is extensive, with a Ph.D. in Global Leadership and Change from Pepperdine University in California, which he is expected to complete in 2023. Sooraj also has two MBA degrees from Pepperdine University, one in Digital Innovation and Information Systems, completed in 2015, and another in General Management, Finance & Marketing, completed in 2012. Additionally, Sooraj holds a bachelor's degree in Electronics & Communication Engineering from C.U.S.A.T in India, which he obtained in 2004.

Sooraj's skills and certifications are diverse and include Analytics Consulting and Analytics Strategy using Data Science/AI/ML, Metrics and Dashboards, Project and Portfolio Management, Statistical Analysis (e.g., GLM, Principal component analysis, discriminant analysis, choice modeling), Lean Six Sigma Black Belt, Certified Scrum Master, Artificial Intelligence, Customer Survey Design, Descriptive Analysis, Inferential Analysis, Feature extraction.

In addition to his impressive professional accomplishments, Sooraj Sushama is also an Adjunct Faculty member at Pepperdine University. He is a respected and dedicated instructor in the MS in Analytics program, sharing his expertise and knowledge with the next generation of data analysts and scientists. Sooraj's commitment to education is a testament to his passion for the field and his desire to inspire and train the next generation of professionals in this growing and exciting industry.

PART I
Blockchain Frameworks

Blockchain Applications and Uses: Deconstructing Blockchain Technology

Blockchain technology provides a sound solution for many industry problems; however, its construct is misconstrued by many. This chapter conducts a comprehensive analysis of introducing relevant content of blockchain technology, provides fundamental principles of blockchain core structure and elements, and underscores opportunities of blockchain technology. Blockchain technology consists of data, the hash of the block itself, and the hash of the previous block. Blocks on the blockchain are created with the hash of previous blocks and added to the infrastructure, therefore the name blockchain. Blockchain technology combines the power of integrated ledger technology that is decentralized, secured, authentic, and establishes trust and integrity. There are four different types of blockchain platforms: private, consortium, public, and hybrid which provide the user with different levels of security. Lack of financial inclusion, availability of big data, and data security problems are prevalent in today's organizations. Blockchain technology works as a distributed ledger that accesses details of all transaction history. The peer-to-peer transmission ensures there is no central authority. Transparency and Pseudonymity are achieved through digital identity. The immutability of records is the consensus of most users to verify transactions. Computational logic is achieved through a digital ledger where algorithms and rules are embedded to trigger automated transactions. If a blockchain tampered with, the rest of the blocks are deemed invalid and the chain breaks. Tampering with one block to make it inactive requires users to perform hashing for all the blocks on the blockchain, redo the proof of work for each block, and the users must have 50% control of the distributed P2P network.

Introduction

The evolution of business is relentlessly concerned with efficiency. Technology has been the lifeblood of business progression. In an increasingly demanding business landscape businesses continually seek innovative techniques for increasing profits, gaining a strategic edge, and creating shareholder value. How do businesses gain strategic advantages in an ever-evolving radically competitive business environment? By understanding and adopting disruptive technologies that pave the way for market differentiation and allow organizations to realize their full potential and strategic advantages for business growth.

This chapter discusses the basics of blockchain technology and allows the reader to understand the platform details and construct of blockchain. This chapter is organized into five sections: introduction, structure, application, conclusion, and chapter summary. First, this chapter presents an overview of blockchain basics in the introduction and provides the fundamental principles of the blockchain structure. Then a focus on the structure and application of blockchain technology to distill key information regarding the technology is displayed. Next, an engaging review of the opportunities for blockchain innovations is provided to conclude the chapter.

Human's ability to make tools cannot be diminished. Blockchain technology is one such tool that is a thrilling advancement in the human toolkit.

Humanity has experienced many evolutionary vicissitudes. The modernizations of the last century have focused on generating value through the growth of information technology. Thus, creating Web 1.0, 2.0, and 3.0 to sort through the massive compilation of available data. In addition, these innovations have cemented discoveries in information technology as the most important invention of humans yet. Through the evolution of the information age, the core struggle of maintaining data integrity has been a challenge. Data privacy has centered on the concept of truth and freedom. ***It is believed that whoever owns the data owns the future or whoever owns the future owns the data.*** However, the massive struggle to streamline and protect data has beleaguered this innovative age of technology. The creation of the World Wide Web offered a brand-new platform that was previously not available to the user. As the popularity grew, businesses and individuals started relying upon the tool and thereby sharing massive amounts of data about their transactions on the platform. Unintended consequences, as is frequently the case with new technology, created deep concerns about data sharing and vulnerabilities surrounding the misuse of data. While the advent of the Web created social and economic inclusion, too much free information on the Internet created a massive problem for data security.

One of the advantages of blockchain technology is its ability to provide secure and transparent transactions. The decentralized nature of this technology eliminates the need for a central authority and provides a high level of security. It also ensures that each party has access to the same information, which promotes transparency and reduces the risk of fraud. Blockchain technology can be applied to a wide range of industries and use cases. It can be used to track the ownership of assets, facilitate cross-border payments, create secure digital identities, and improve supply chain management, among other things. The technology has the potential to create a more inclusive financial system by providing access to financial services to those who are currently excluded.

There are also challenges and limitations to the widespread adoption of blockchain technology. For example, this technology is still in its early stages, and there is a lack of standardization and interoperability across different blockchain platforms. Additionally, there are concerns about the scalability and energy consumption of blockchain networks. Blockchain technology has the potential to revolutionize the way we conduct transactions and manage data. Its core principles of decentralization, security, transparency, and immutability make it an attractive solution for a wide range of industries and use cases. However, there are also challenges and limitations to the widespread adoption of the technology that need to be addressed.

The next generation of technology is seeking a solution for data security. How do we make value out of the free data floating around on the internet? The internet redefined how we share information but the lack of autonomy and accountability in sharing the data is where the value is created. It can be said that we are headed towards an age of autonomy or the value age. As the internet redefined sharing of data, blockchain technology is redefining how we transact. What we need to understand is that the core values of the blockchain technology foundation are autonomy (trust) and accountability (freedom). The information floating around on the internet can be better understood as a digital asset. What is a digital asset? A digital asset encompasses resources, rights such as voting rights, collectibles, records, or utility. The information age has pushed how our civilization operates. Everything of existence or value will be transformed into blockchain and will live on the blockchain in the future. All internet or databases will be on the decentralized web utilizing blockchain technology. How does blockchain achieve efficiency, effectiveness, and accountability? By streamlining processes in business, securing data, and empowering individuals by saving time and money.

Blockchain technology achieves efficiency, effectiveness, and accountability by providing a decentralized and transparent system of record-keeping. The use of blockchain technology eliminates the need for intermediaries such as banks or government agencies in transactions, which reduces transaction costs, eliminates the potential for fraud, and ensures accountability. Blockchain technology uses cryptography to ensure the security and privacy of data. Each transaction is recorded in a block that is linked to the previous block, creating a chain of blocks, hence the name blockchain. Once a block is added to the blockchain, it cannot be altered, which provides a tamper-proof record of all transactions.

Blockchain technology also enables the creation of digital assets that can be traded or used in various ways. Digital assets are stored on the blockchain, making them immutable, secure, and easily traceable. This creates a new era of ownership where people can own and trade their digital assets with greater freedom and autonomy. Furthermore, blockchain technology provides a new paradigm of autonomy and accountability in sharing data and transacting, creating a new era of value creation. It offers many advantages, including security, privacy, and transparency, which can improve business processes, save time and money, and empower individuals.

Data security is a multibillion-dollar business. Advancements in technology have created rapid computing devices that enable perpetrators to create fraud that costs businesses monetary and reputational damages. How must businesses protect their data, processes, and risk of reputational damage by utilizing blockchain technology? While the rapid evolution of the information data age creates severe vulnerabilities, it also creates vast opportunities for the next generation of technologies.

Blockchain technology can be a valuable tool for businesses to protect their data, and processes, and risk reputational damage. Blockchain is a decentralized, immutable ledger that allows for secure and transparent data sharing and storage. Here are some ways businesses can use blockchain technology to enhance their data security:

1. **Secure data storage:** Blockchain technology can provide a secure and tamper-proof platform for businesses to store their sensitive data. The decentralized nature of blockchain means there is no single point of failure, making it extremely difficult for hackers to breach the system.
2. **Improved data sharing:** Businesses can use blockchain technology to securely share data with partners and customers. The transparent nature of blockchain ensures that all parties have access to the same information, reducing the risk of miscommunication and misunderstandings.
3. **Smart contracts:** Smart contracts are self-executing contracts that automatically enforce the terms of an agreement. Businesses can use smart contracts to automate processes and reduce the risk of fraud.
4. **Identity verification:** Blockchain technology can be used for secure identity verification, reducing the risk of fraud and identity theft.
5. **Reputation management:** Blockchain can be used to track and verify the authenticity of a business's products or services, enhancing its reputation and reducing the risk of reputational damage.

Blockchain technology provides a secure and transparent platform for businesses to protect their data, processes, and reputation. By leveraging blockchain, businesses can enhance their data security and reduce the risk of monetary and reputational damages.

Purpose: This chapter aims to provide a toolkit to those who do not understand the basics of blockchain and how it can be applied in a real-world application.

Potential Solution: What is the resolution?

1. To clearly define blockchain
2. To break down barriers of complexity by unraveling the basic composition of blockchain technology

3. To provide clear frameworks that show the practical application of blockchain technology in various industries.

What is a Blockchain?

A chain of blocks that contains information. Blockchain technology is an intimidating, bewildering, vague, and multifaceted technology that is greatly misinterpreted.

Timeline: The first use of blockchain technology was introduced in 1991 for timestamping purposes. The concept of the technology was to avoid alterations in digital documents. In 2008, Satoshi Nakamoto created Bitcoin—a cryptocurrency that brought blockchain technology to the forefront of innovative measures.

Blockchain Basics

Blockchain technology is the newest phase of innovative technology advancement that is still largely misunderstood. Business leaders seldom understand the complexities that underpin this disruptive and complex technology. Lack of alacrity regarding how blockchain works lends to more confusion regarding the technology and creates perplexities for leaders looking for new technological solutions. This book lends itself as a practitioner/application tool for those who seek to understand the framework of blockchain technology, its benefits, and its applications. The authors have carefully crafted sections of the book that are dedicated to different industry's blockchain applications and uses in the financial industry, healthcare industry, law enforcement industry, and supply chain industry. Individuals who seek clarity and an in-depth understanding of how blockchain technology is utilized in these fields will greatly benefit from the content and detailed frameworks shaped to unpack confusion regarding blockchain technology. Blockchain technology is still widely misunderstood, and there is a need for more education and awareness around its applications and benefits. This book will serve as a valuable resource for anyone seeking a deeper understanding of blockchain technology. By providing practical examples and detailed frameworks, the authors aim to help readers better grasp the complexities of blockchain technology and how it can be applied to solve real-world problems. This is particularly important given the potential impact that blockchain technology can have on various industries, including improved transparency, efficiency, and security. This book could be a useful tool for business leaders, entrepreneurs, and others seeking to gain a better understanding of blockchain technology and its potential applications.

Organizations are constantly seeking technological advancements to reform legacy systems and step into the future, and misunderstanding blockchain technology is presently a missed opportunity in many organizations. Zehir et al. (2022) identified that corporate executives have not been successful in identifying the correct type of blockchain platform for their sectors. Blockchain technology allows enhanced data management, reduces transaction costs, and avoids fraud and human error by facilitating performance and enhancing monitoring, and maintenance of the data. Furthermore, the integrated use of Artificial Intelligence (AI) and blockchain technology is another solution to resolve design and integration issues for organizations seeking to explore blockchain technology. Blockchain technology is a customizable approach that allows users to benefit from its different forms of approach such as characteristics, consensus mechanisms, and varieties of designs that are customizable. Additionally, blockchain provides a distributed digital ledger through a decentralized system (Saldamli et al., 2022) and maintains data integrity, consistency, and transparency through a robust distributed huge network.

Blockchain is a reservoir for storing large data sets of information. Blockchain provides a framework for decentralized operations enabling users to communicate securely and directly. This also facilitates the storage of information as digital signatures that allow financial transactions to be executed efficiently. In addition, blockchain technology utilizes security, authenticity, and integrity to establish trust. The data in a blockchain is a ledger or list of records (blocks) that are securely interconnected. For instance, the financial industry constantly battles consumer trust. How can blockchain technology resolve the problem of trust? Blockchain uses the mathematical principles of measurement that follow the premise of traceability. The fundamentals of mathematical measurement occur through a documented unbroken chain of calibrations that are like blockchains (Milicevic et al., 2022). Blockchain uses a similar premise for calibration that supports measurement authenticity for each block, thereby, confirming and building traceability and trust. Furthermore, smart contracts benefit from measurement traceability. Smart contracts are based upon the premise of defined terms and conditions. When terms and conditions are met the smart contract automatically executes.

Blockchain technology can help resolve the problem of trust in the financial industry, as well as in other industries where trust is important. Blockchain achieves this by providing a tamper-proof, decentralized ledger of transactions that cannot be altered or manipulated without the consensus of the network. Traditionally, financial transactions are processed by a centralized authority such as a bank, which acts as a trusted intermediary between parties. However, this system is vulnerable to fraud, corruption, and human error, which can erode trust in the system. With blockchain technology, transactions are verified and recorded by a distributed network of computers, and once recorded, the transactions cannot be altered or deleted. This creates a transparent and tamper-proof system that can be trusted by all parties. The use of digital signatures in blockchain technology allows for the secure and efficient execution of financial transactions without the need for intermediaries, further reducing the potential for fraud or error. Therefore, blockchain technology has the potential to significantly improve trust in various industries by providing a transparent, tamper-proof, and decentralized system for storing and verifying information.

Blockchain utilizes a multilayer approach that is distributed into five categories: network layer, consensus layer, data model layer, execution layer, and application layer (Kaur et al., 2022). Blockchain technology is an innovative option for transferring operations in various industries. Blockchain is a peer-to-peer network of nodes that facilitates communication across a network of non-trusting nodes and adds new blocks to the end of an existing block without altering the existing previous block.

Benefits of consensus:

1. Consensus determines the network security and performance to ensure the long-term stabilization of the blockchain.
2. Consensus ensures that trusted intermediaries maintain control through built-in algorithms that validate and confirm every transaction of the blockchain.

A chain of blocks that contains information (1991) that are timestamp-concept generated from digital stamps concepts to avoid alteration (2009). Bitcoin was created by Satoshi Nakamoto as a form of digital currency. Blockchain technology is a distributed ledger that means access to all and is completely open to everyone. Once data is recorded on a blockchain it is extremely difficult to alter.

Blockchain consists of:

1. Data
2. The hash of the (block itself)
3. The hash of the previous block

The data is where the use of technology lies for various industries. For example, in a financial transaction done through Zelle type of technology, there are key pieces of information that maintain the record: From, to, amount in the stored data. Hash is the unique fingerprint; therefore, if the hash changes, the information is no longer the same. Proof of work is required to create a new block.

Block creation: Genesis block is the first block that is created and does not have a previous hash.

Figure 1 shows how blockchain blocks are created and highlight the unique identifiers that are attached to each block as the hash of the previous block. **Note:** The genesis block does not have a previous hash because it is the first one created.

As shown in Fig. 1, the genesis block is the first block that is created; therefore, the genesis block does not have a previous hash and a unique identifier (the hash of the block itself) is assigned to the genesis block. However, as the next block is created, the new block is assigned a new hash identifying itself called the hash of the block and note that the hash of the previous block matches the genesis block. If one block in the blockchain is tampered with, it will make the rest of the blocks invalid as the chain breaks. Computers' computing power creates an environment where all the hashing can be reconfigured and recalculated easily to make a block valid again (invalid block) tampered block to mitigate this (fast computing) slows down the creation of new blocks. Tamper with one block

**Genesis
Block**

Hash: 125Q
Previous Hash: 0000

Hash: 569L
Previous Hash: 125Q

Hash: 942P
Previous Hash: 569L

Figure 1: Blockchain creation (Source: Author's own work)

means that now the system must reconfigure the proof of work for each of the following blocks on the blockchain. To tamper with one block on the blockchain, one will have to do the following.

To tamper with blockchain:

1. All the blocks of a blockchain.
2. Redo the proof of work for each block.
3. And take control of 50% of the P2P network.

When new blocks are created, nodes are verified and then added to the blockchain. Each blockchain participant validates the blocks = verified through consensus. Only then the new block will be accepted by all users as valid. The basic construction of a blockchain is dependent upon the following:

1. Hashing
2. Proof of work
3. Distributed P2P networks (everyone gets a copy of the same blockchain)

Blockchain technology is a decentralized technology that provides transparency through the distribution of ledger databases. Rapid digital development of new generational technology has realized development better and seeks to achieve effectiveness in business transformation. The future of smart enterprise continues to seek the development of technology and is the driving force of enterprise information. Businesses seek efficient operations systems that can organize data and information to analyze data and provide decision-makers with information to create market competencies (Zhu, 2022). Blockchain technology uses a distributed node network system where consensus is at its core. There is no administrator, and all nodes contain data. Blockchain networks follows the same rules as consensus in return consensus creates new blocks for the network. The consensus mechanism is supported by algorithms that are reached through the sequence of transactions in a time window and consensus determines the level of trust between the nodes of the blockchain. Blockchain technology is a robust solution for data security. Therefore, many industry problems that can be solved using blockchain-based solutions. Here are some important issues that can be resolved using blockchain.

What are the problems industries are currently facing? What are the pain points?

1. Heavy regulations
2. Bad actors
3. Consumer lack of trust
4. Fragmentation
5. Industry lingo
6. The industry needs to protect its own data
7. The industry needs to protect consumer data
8. Consumer data sharing across different divisions to drive value for clients and the firm.

Corporate executives have not successfully identified the correct type of blockchain platform for their sectors. Blockchain technology allows enhanced data management, reduces transaction costs, and avoids fraud and human error by facilitating performance and enhancing, monitoring, and maintaining the data. The integrated use of AI and blockchain technology is yet another solution to resolve design and integration issues for organizations seeking to explore blockchain technology (Zehir et al., 2022). Blockchain technology is a customizable approach that allows users to benefit from its different forms of approach such as characteristics, consensus mechanisms, and varieties of designs that are customizable. There are **four** types of blockchains.

Types of blockchain technologies:

1. Private
2. Consortium
3. Public
4. Hybrid

The use of blockchain depends upon the needs of an organization. Private blockchain grants access to only approved participants by exercising a single authority. A consortium blockchain is controlled by a group and not a single authority. A public blockchain is accessible to anyone. A hybrid blockchain is a solution that uses a customizable approach where members choose who can join the network and what information is open to the public. As organizations decide on which type of blockchain to consider, some risks are also associated with the decision-making practice.

Organizations are actively seeking to minimize risks. There are **eight** types of risk that firms should consider: strategic risk, business continuity risk, reputational risk, information security risk, regulatory risk, operational risk, contractual risk, and supplier risk.

1. Strategic risk is a corporation's decision to onboard technology in the early business phase or to wait until the business is mature.
2. Business continuity risks are service interruptions due to cyber attack that is directly related to business continuity risk.
3. Reputational risk is associated with compliance with legacy systems.
4. Information security risk is related to consumer data stored in private blockchains.
5. Regulatory risk is associated with effectively meeting the compliance guidelines of various regulatory agencies.
6. Operational risk is related to the scalability, speed, and inoperability of legacy systems.
7. Contractual risk is associated with agreements between participating nodes and network administration.
8. Supplier risks are a product of risks that are associated with 3rd party providers.

Blockchain application to solve the world's most pressing issues has been a proven success in many industries. Industries that desire streamlined processes yet are fragmented in different sectors can gain information transparency by applying blockchain frameworks. Blockchain evolution has occurred in different phases and although the technology is relatively new, it is currently in Phase 3, where full integration of the technology is commencing in various industries.

Blockchain Phases
1. Phase 1 – 1.0 era that started with the advent of digital currency.
2. Phase 2 – 2.0 era added the use of smart contracts.
3. Phase 3 – 3.0 era is fully integrating applications.

The application and benefits of blockchain technology in different industries have decision-makers globally realizing the huge potential of blockchain technology and its capacity for problem-solving, transparency, and removing obstacles such as information opacity, and high transaction costs by providing solutions for data security (Zhao et al., 2022). Industries have actively searched for a solution surrounding the protection, integration, and sharing of data. Although many information technology solutions are currently available on the market, blockchain-based solutions lead the revolution.

Structure

The opportunity for blockchain technology to include financial inclusion is tremendous. Over two billion people have limited or no access to financial services in developing countries worldwide (Larios-Hernandez, 2017). Paradoxically, the world of financial services has greatly benefited from blockchain-based technology adoption. According to many authors, bitcoin is the largest implementation of blockchain technology and is successful in overcoming challenges related to infrastructure, scalability, inflexibility, and unsustainability (Back et al., 2014; Eyal et al., 2016). Even though the popularity of Bitcoin has put blockchain technology at the forefront of technological evolution, the technology is a robust solution for many industries. Blockchain technology opportunities stem from the technology's capacity to operate on nodes that grant access to active users, an issue that incumbent systems have been unable to accomplish. George et al. (2012) assert that blockchain creates opportunities that augment the social and economic well-being of marginalized society members.

Blockchain technology has the potential to greatly improve financial inclusion in developing countries. The decentralized nature of blockchain allows for transactions to be processed without the need for a centralized intermediary, such as a bank. This makes it easier for individuals who may not have access to traditional financial institutions to participate in the global economy. In addition to financial inclusion, blockchain technology also has the potential to improve transparency and reduce corruption in industries such as supply chain management and voting systems. The immutable nature of blockchain means that once a transaction is recorded, it cannot be altered or deleted, providing a tamper-proof record of events. It is important to note that blockchain technology is not a panacea and there are still challenges to be addressed, such as scalability, interoperability, and user adoption. Moreover, the environmental impact of blockchain mining is a growing concern that needs to be addressed. While blockchain technology has the potential to bring about significant benefits to society, it is important to approach its implementation with a critical eye and consider both the potential benefits and drawbacks.

The availability of big data companies and social networks already supports the success of blockchain technology in facilitating financial transactions in underserved communities. The financial industry has historically used a top-down approach that fails to organize the local contexts and practices of unbaked individuals. Dahan and Gelb (2015) assert that there are over 2.4 billion persons around the world who do not have a legal identity. Among other issues, lack of identity is a problem in gaining access to financial institutions. An informal social capital structure has been prevalent in supporting the financial needs of unbanked individuals. Establishing blockchain technology creates safer transactions, especially for the informal social capital structure that has previously only relied upon relationships to fund financial needs for individuals without access to traditional banking. Therefore, reputation-based peer-to-peer lending services create future opportunities for blockchain technology. There is a huge unmet need for unbanked world citizens that creates a tremendous opportunity for blockchain-based technologies to create a financial network such as microcredit to encourage participation. Strategic implementation of blockchain-based systems to transform the traditional banking system is much needed. There are many advantages for financial platforms to consider blockchain-based architecture to serve the world's population that does not have access to traditional banking.

Application

Blockchain technology has the potential to revolutionize existing infrastructure and garner opportunities for developing nations (Gupta and Knight, 2017). A detailed construction of blockchain network architecture is shown in Fig. 2. The three layers of architecture include the data storage layer, central management layer, and user layer (blockchain layer). In the data storage layer, all databases that are currently being utilized are included, for instance, shared databases, local databases, and any databases of additional records used by the organization. In the central management layer, certificate of authority, transaction logs, etc. are included, where the actual transaction gets validated and authenticated. The user layer (proposed blockchain layer) is where the data is ultimately displayed to the user in near real-time. To further drill down into the blockchain mechanism, we have illustrated five steps of the blockchain blueprint.

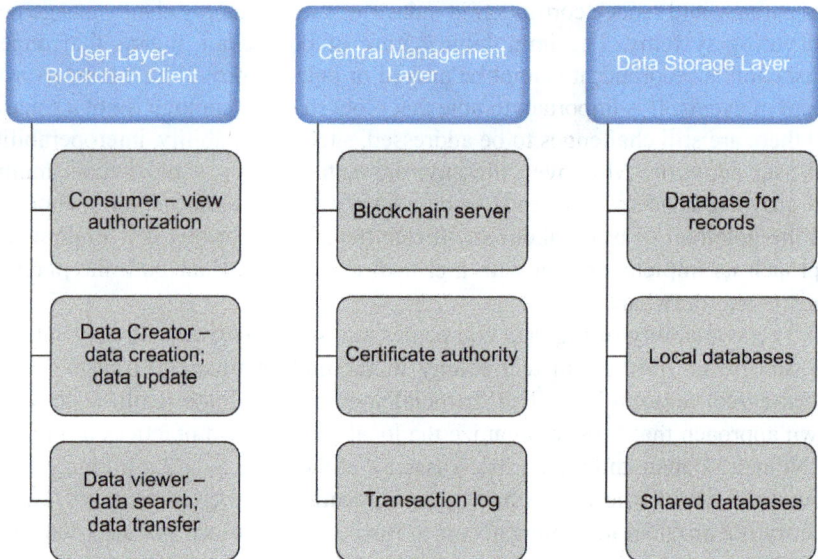

Figure 2: Blockchain network architecture (Source: Author's own work)

These five steps are repeated for every transaction occurring on the blockchain, ensuring that the ledger remains transparent, secure, and immutable. By leveraging the power of blockchain technology, organizations can build decentralized, trustless networks that enable secure and efficient transactions without the need for intermediaries.

Step 1: Transactions – The first step in the blockchain mechanism involves the initiation of a transaction. This can be any type of transaction, such as a payment or the transfer of assets, and is recorded in a block.

Step 2: Verification – Once a transaction is initiated, it needs to be verified by the network. This is done by a network of nodes (computers) that work together to validate the transaction using complex algorithms.

Step 3: Block creation – Once the transaction is verified, it is added to a block along with other validated transactions. Each block contains a unique digital signature, a timestamp, and a reference to the previous block in the chain.

Step 4: Block validation – Before a block can be added to the blockchain, it needs to be validated by the network. This is done by solving a complex mathematical problem that ensures that the block is legitimate.

Step 5: Blockchain confirmation – Once a block has been validated, it is added to the blockchain and becomes a permanent part of the ledger. The blockchain is then replicated across all nodes in the network, ensuring that every participant has an up-to-date copy of the ledger.

Blockchain Network Architecture

How Blockchain Works: Breaking down the technology into **five** basic steps.

1. **Distributed Ledger:** Every user on the blockchain has access to the entire blockchain. The access details of all transaction details and complete history. There is no single-party control of the data, and every user has access to verifying the records of other users directly without the intervention of an intermediary.
2. **Peer-to-Peer Transmission:** There is no central authority. All communication happens directly between the users.
3. **Transparency and Pseudonymity:** All transactions are visible to anyone who has access to the blockchain platform. Digital identity is built in for each user and there are alphanumeric identifiers that allow users to remain anonymous.
4. **Immutability of Records:** Once a transaction is recorded on the blockchain system, any alteration of records signals updates to previous records. Most users on the blockchain must then verify the validity of that changed transaction. Thus, alterations on the blockchain are highly visible and controlled by a majority and cannot be easily altered.
5. **Computational Logic:** As a digital ledger, blockchain transactions can be related to computational logic that is embedded in algorithms and rules that trigger automated transactions between nodes on a blockchain.

An early example of blockchain technology facilitating leapfrogging in underdeveloped nations is M-Pesa (M for mobile and Pesa is Swahili for money). M-Pesa is a first-of-its-kind mobile phone-based money transfer service that originated back in 2007 in Kenya. The lack of a traditional banking structure created burdensome money transfer transactions. The use of blockchain technology to facilitate payments to and from poor farmers changed the lives of thousands from unbanked transactions to successful money inflows and outflows. Banking infrastructures in many countries are extremely limited to the population. This means that the general population in these countries is locked out of the banking system. Blockchain creates a financial system where formal banking is transformed into mobile banking reducing the miles of distance traditionally created in the past. Mobile banking does not require human intervention or a brick-and-mortar banking facility.

Lack of banking infrastructure and consumer habits in developing economies have generated cash hoarding. Mobile money exchange has been a positive step toward profound growth. The application of blockchain-based technology creates transactional security that extends beyond biometrics which only secures the related transaction, whereas blockchain could secure the entire system and transactional process. Developing nations are excellent benefactors of blockchain technology as the functions of the technology provide natural decentralization meaning that distance to infrastructures and data centers does not matter to the consumer. The modernization and digitalization of government functions in developing countries can also benefit from the application of blockchain technology. For instance, Dubai Blockchain Strategy was enacted to move all government documents onto a blockchain platform to create huge cost savings and efficiencies. The Dubai Blockchain Strategy has resulted in a collaboration between many government agencies to explore, evaluate, and deliver the best opportunities for its citizens. The initiative has already created economic opportunities and digital innovations. It has been cited that the adoption of blockchain technology unlocks 5.5 billion dirhams in savings annually, just in document processing alone compared to the total value of the Burj Khalifa.

Alongside governments such as the Dubai example above, famous economists realize the value of this futuristic, innovative technology. Economist Hernando de Soto has dedicated his career to improving access to the formal economy and believes that the reason underprivileged individuals don't have more access to the formal economy is twofold: most recordkeeping systems are unreliable and lack of trust between the individual and the organization – simply put individuals are not willing to give up their information because they do not trust organizations and government. However, blockchains' tamperproof and autonomous features are very attractive to eradicating the lack of access to the world's poor population.

Hernando de Soto is a renowned economist who has spent his career working to improve access to the formal economy for underprivileged individuals. He believes that one of the main reasons why many people lack access to the formal economy is due to unreliable record-keeping systems and a lack of trust between individuals and organizations. De Soto sees the potential for blockchain technology to help address these issues by providing a tamper-proof and autonomous system for recording and verifying transactions. The transparency and immutability of blockchain technology could help to build trust between individuals and organizations, potentially making it easier for underprivileged individuals to participate in the formal economy.

In addition to de Soto, many other economists and experts see the potential for blockchain technology to revolutionize various industries and promote financial inclusion for underprivileged populations. By leveraging the unique features of blockchain technology, such as decentralization, immutability, and transparency, organizations can create new and innovative solutions to longstanding problems in fields like finance, supply chain management, and more.

Conclusion

Blockchain technology is a chain of blocks that contains information. Blockchain technology is an intimidating, bewildering, vague, and multifaceted technology that is greatly misinterpreted. It was first introduced in 1991 for timestamping purposes. The concept of timestamping technology was to avoid alterations in digital documents. In 2008, Satoshi Nakamoto created Bitcoin – a cryptocurrency that brought blockchain technology to the forefront of innovative measures. Organizations are constantly seeking technological advancements to reform legacy systems and step into the future, and misunderstanding blockchain technology is presently a missed opportunity in many organizations.

Blockchain technology is essentially a decentralized, distributed ledger that enables secure, transparent, and tamper-proof transactions. It works by using cryptographic algorithms to create a digital signature for each block of data, which is then added to the chain of previous blocks in chronological order. The result is an immutable record of transactions that can be verified by anyone in the network.

The potential applications of blockchain technology are vast, ranging from financial services to supply chain management, healthcare, voting, and more. It has the potential to eliminate intermediaries, increase transparency and security, and reduce transaction costs.

However, there are still many misconceptions and misunderstandings about blockchain technology. Some people believe it is synonymous with cryptocurrency or a magic solution to all problems. Others are intimidated by the complex technical jargon or skeptical of its potential. It's important to educate oneself about blockchain technology and its potential applications to make informed decisions about its use. While it may not be suitable for every organization or use case, it is certainly worth exploring as a potential solution to some of the challenges faced by businesses and industries today.

Blockchain utilizes a multilayer approach that is distributed into five categories: network layer, consensus layer, data model layer, execution layer, and application layer. The benefits of consensus are:

- Consensus determines the network security and performance to ensure the long-term stabilization of the blockchain.
- Consensus ensures that trusted intermediaries maintain control through built-in algorithms that validate and confirm every transaction of the blockchain.

The architecture of Blockchain consists of

- Data
- Hash of the (block itself)
- Hash of the previous block

There are **four** types of blockchains

1. Private
2. Consortium
3. Public
4. Hybrid

As organizations decide on which type of blockchain to consider, some risks are also associated with the decision-making practice. Organizations are actively seeking to minimize risks. There are eight types of risk that firms should consider: strategic risk, business continuity risk, reputational risk, information security risk, regulatory risk, operational risk, contractual risk, and supplier risk.

As organizations consider implementing blockchain technology, it is important to be aware of the various risks associated with the decision-making process. Here is a brief overview of the **eight types of risk** that firms should consider:

1. **Strategic risk:** This refers to the risk that blockchain implementation may not align with the organization's overall strategy and objectives.
2. **Business continuity risk:** This refers to the risk that blockchain implementation may disrupt business operations or result in the loss of critical data or services.
3. **Reputational risk:** This refers to the risk that blockchain implementation may result in negative publicity, damage to the organization's brand, or loss of customer trust.
4. **Information security risk:** This refers to the risk that blockchain implementation may result in security breaches, data loss, or unauthorized access to sensitive information.
5. **Regulatory risk:** This refers to the risk that blockchain implementation may not comply with applicable laws and regulations, resulting in legal or financial penalties.
6. **Operational risk:** This refers to the risk that blockchain implementation may not function as intended, resulting in errors, delays, or other operational issues.
7. **Contractual risk:** This refers to the risk that blockchain implementation may not meet contractual obligations or result in disputes with business partners or vendors.
8. **Supplier risk:** This refers to the risk that blockchain implementation may rely on third-party suppliers, who may not meet quality, reliability, or security standards.

By being aware of these risks and implementing appropriate risk management strategies, organizations can minimize the potential negative impacts of blockchain implementation and ensure the successful integration of the technology into their operations.

Blockchain evolution has occurred in different phases and although the technology is relatively new, it is currently in Phase 3, where full integration of the technology is commencing in various industries.

Blockchain Phases

There are 3 phases associated with Blockchain. Phase 1 is the first phase where digital currency, such as, Bitcoin was created on the blockchain framework.

Phase 2 is the second phase where smart contracts were integrated into the blockchain technology via the use of algorithms to manage logic statements to satisfy contract limits.

Phase 3 is the third phase of blockchain technology where a fully integrated blockchain framework is experienced.

How Blockchain Works: Blockchain technology can be broken down into five steps.

1. Step 1 is the distributed ledger where blockchain technology utilizes a distributed ledger where all participants in the network have a copy of the entire blockchain.
2. Step 2 is peer-to-peer transmission, which means no central authority or intermediary is involved in the transaction process.
3. Step 3 is transparency and pseudonymity which is a fundamental characteristic of blockchain. All transactions are visible to participants who have access to the blockchain.
4. Step 4 is the immutability of records, which means it is extremely difficult to alter once a transaction is recorded on the blockchain.
5. Step 5 is computational logic that allows for the implementation of smart contracts through programmable logic.

Blockchain technology is a powerhouse that can resolve big data sharing, integration, and security problems for many industries and leaders must thoughtfully consider its application.

Chapter Summary

Blockchain technology works as a distributed ledger that accesses details of all transaction history in real time to all participants on the blockchain. Blockchain technology utilizes security, authenticity, and integrity to establish trust. The data in a blockchain is a ledger or list of records (blocks) that are securely interconnected. To establish trust, a Blockchain uses mathematical principles of measurement that follow the premise of traceability. Blockchain utilizes a multilayer approach that is distributed in five categories: network layer, consensus layer, data model layer, execution layer, and application layer (Kaur et al., 2022). Blockchain technology is an innovative option for transferring operations in various industries. Blockchain is a peer-to-peer network of nodes that facilitates communication across a network of non-trusting nodes and adds new blocks to the end of an existing block without altering the existing previous block.

The peer-to-peer transmission ensures there is no central authority. Transparency and Pseudonymity are achieved through digital identity. The immutability of records is the consensus of most users to verify transactions. Computational logic is achieved through a digital ledger where algorithms and rules are embedded to trigger automated transactions. If a blockchain has been tampered with the rest of the blocks are deemed invalid and the chain breaks. Tampering with one block to make it inactive requires users to perform hashing for all the blocks on the blockchain, redo the proof of work for each block, and the users must have 50% control of the distributed P2P network. Blockchain consists of the data hash of the (block itself) and the hash of the previous block. The data is where the use of technology lies for various industries. Block creation: genesis block is the first block that is created and does not have a previous hash.

If one block in the blockchain is tampered with, it will make the rest of the blocks invalid as the chain breaks. The computer's computing power creates an environment where all the hashing can be reconfigured and recalculated easily to make a block valid again (invalid block). Tampered block to mitigate this (fast computing) slows down the creation of new blocks. Tamper with one block now and you must reconfigure the proof of work for each of the following blocks on the blockchain. In order, one will have to do the following: Hashing, proof of work, and distributed P2P networks (everyone gets a copy of the same blockchain). When new blocks are created, nodes are verified and then added to the blockchain. Each blockchain participant validates the blocks = verified through consensus. Only then the new block will be accepted by all users as valid.

Take Away Lessons from the Chapter

1. Blockchain technology is a revolutionary, disruptive yet viable technology. It consists of data, the hash of the block itself, and the hash of the previous block. Blocks on the blockchain are created with the hash of previous blocks and added to the infrastructure, therefore the name blockchain.
2. Blockchain technology combines the power of integrated ledger technology that is decentralized, secured, and authentic and establishes trust and integrity.
3. There are four different types of blockchain platforms: private, consortium, public, and hybrid which provide the user with different levels of security.
4. There are three blockchain phases: Phase 1 – 1.0 era that started with the advent of digital currency; Phase 2 – 2.0 era added the use of smart contracts; and Phase 3 -3.0 era is fully integrating applications.

5. Lack of financial inclusion, availability of big data, and data security problems are prevalent in today's organizations. Blockchain technology works as a distributed ledger that accesses transaction history details.
6. Organizations are actively seeking to minimize risks. There are eight types of risk that firms should consider: strategic risk, business continuity risk, reputational risk, information security risk, regulatory risk, operational risk, contractual risk, and supplier risk.
7. Blockchain technology works as a distributed ledger that accesses transaction history details.

Chapter Exercises

1. Form student teams of three to five. Strive for teams to consist of individuals from different industries. To pre-assess the understanding of blockchain technology, have the students break into their respective groups and craft an in-class message. For example, what do you think of blockchain technology? Or can you describe blockchain technology? How do you think blockchain is relevant or not relevant in your industry? After the in-class-exercise, have students report on their findings.
2. Consider a list of advantages of blockchain technology in your industry.
3. What barriers have you found to be most significant in onboarding new technology such as blockchain? What can be done to overcome or minimize these barriers?

Blockchain Applications and Uses: Finance Industry

The finance industry has been one-sided since inventing blockchain technology for fiat (fiat currency is not backed by a commodity such as gold and silver), and cryptocurrency. Although the advent of blockchain technology took place in 2008 with the introduction of Bitcoin, the technology is greatly misunderstood as a facilitator of cryptocurrency-based transactions only. This fallacy must be resolved. While digital currency is very widespread and deeply associated with blockchain technology, the technology is useful in all aspects of the finance industry. This chapter focuses on the lack of regulations surrounding digital assets that threaten the global financial sector. It delves into the challenges that lawmakers currently face to properly enforce the regulation of digital assets. Then, how blockchain can be an effective solution in eradicating digital asset exploitation that includes cybercriminal activities, money laundering, and terrorist financing is discussed. The impact of financial crimes on economies is outlined, and a conceptual framework is constructed by the authors to show how blockchain technology would theoretically and practically operate in the finance industry. The construct of the framework is divided into three sections: the physical layer, the current layer, and the blockchain layer for simplifying the integration of blockchain technology in the finance sector. Next, the application of blockchain technology in the finance industry is discussed in detail with the inclusion of rich case studies to support the benefits of blockchain technology in the finance sector. The chapter concludes with thoughtful highlights of the chapter and an inclusive exercise to develop further thoughts on blockchain technology.

Introduction

The increasing hype of digital currency is persistently shifting and transcending national borders and keeps its user's identity anonymous. While digital currency is widespread it has been afflicted by illegal activity forcing law enforcement to question its effectiveness. Although the Federal government has intervened in several legal and regulatory matters regarding digital currency and fraud, there is a substantial ever-expanding gap between digital currency and organized crime. The application of the law has been slow-moving and therefore has created challenges. Digital currency's main characteristics are decentralization and anonymity. Criminal activities are pervasive and harmful to all industry participants. Activities such as hacking, ransomware attacks, money laundering, counterfeiting, and drug trafficking are some illegal activities that have adopted cryptocurrencies to circumvent the law (Mikulic, 2020). Law enforcement faces a precipitous challenge when digital assets are used by criminals to purchase drugs, weapons, and other contraband anonymously. Digital currency is stored in a wallet. A digital wallet is a secure physical device or software program that stores, transfers and receives digital currency (Pieroni, 2018). The increase in digital currency use for financial crimes is rising. The Australian government recently implemented a crackdown on drug trafficking by exposing the relationship between organized crime activities and digital currency (Pieroni, 2018). Apart from money laundering, digital assets are often used to fund terrorist activities. Counter-terrorism experts have well-documented the use of digital assets to fund terrorism. According to Paul (2018), in April 2017, the Islamic State of Iraq and Syria (ISIS) used the dark web to raise approximately $3 million in digital assets to secure funding for their terror activities. Moreover Akyildirim et al. (2020) identified that in 2020 the world economic forum predicted that the online crime market value can increase to more than $2 trillion due to the growth in digital currencies.

There is an absence of checks and balances that harbors cybercriminals and facilitates illegal activities. Digital assets are linked to global money flow without limitations. Digital assets work on the same premise as blockchain technology. Digital assets provide buyers and sellers with transparency and security; however, the lack of government enforcement keeps law enforcement in the dark (Hughes and Brown, 2022). Digital assets can easily be tracked and all users that share the blockchain to store the digital assets can be shared with law enforcement to provide transparency and visibility into illegal activities conducted by unscrupulous participants. The lack of regulations on virtual assets has created a significant burden on law enforcement, diminished investor protection compared to conventional securities markets, and proliferated fraud and manipulation in the financial sector (Clayton, 2017). According to Turner and Irwin (2018), over 5,000 cryptocurrencies are trading in over 200 countries and the market is growing at an exponential rate. With the advent of new cryptocurrency platforms, there are more concerns arising within the financial sector because of the inherent misuse of this digital asset to launder money, finance terrorism and evade taxes.

Digital assets, such as cryptocurrencies, are indeed a relatively new concept that poses a challenge for regulators and law enforcement agencies worldwide. While they offer many benefits, such as increased transparency and security, the lack of proper regulations and enforcement can lead to their misuse and facilitate illegal activities. Cryptocurrencies have been associated with money laundering, terrorism financing, and tax evasion due to their anonymous nature and lack of oversight (Ali et al., 2020). In response, many governments and regulatory bodies are starting to take action to mitigate these risks by implementing laws and regulations that would increase transparency, protect investors, and deter illegal activities. However, regulating digital assets is a complex task that requires balancing innovation and development with protecting the public interest. Therefore, it is crucial to find the right balance between protecting investors and preventing financial crimes while still allowing for innovation and development in the digital asset industry. The lack of proper regulations and enforcement on digital assets poses a challenge for law enforcement agencies and regulators (De, 2022). While digital assets offer many benefits, their misuse can facilitate illegal activities. Consequently, governments and regulatory bodies must find the right balance between protecting investors and preventing financial crimes while still allowing for innovation and development in the digital asset industry.

The lack of regulations surrounding digital assets threatens the global financial sector. Lawmakers are presented with a challenging task regarding enacting new regulations for digital assets (Breu and Seitz, 2018). According to Breu and Seitz, lawmakers will have difficulty enforcing criteria and the self-governing nature of the digital asset storage system inherently forbids the influence or control of a central authority. While digital currency is largely self-regulated today, governance needs to be developed around the mechanism to control the cash flow from cryptocurrencies, which would not be a violation of the technology. However, currently, there is a lack of comprehensive compliance programs that include compliance policies, customer identification, due diligence, and reporting. Current regulations in the banking industry would simply apply to the tracking and maintenance of digital currency transactions; however, lawmakers need to define and add regulatory policies to digital assets so law enforcement agencies can effectively enforce them.

The lack of regulations surrounding digital assets indeed presents a challenge for the global financial sector. As digital assets become more prevalent, it is crucial to have a regulatory framework to ensure their safe and secure use. The decentralized nature of digital asset storage systems poses a significant obstacle for lawmakers in enacting regulations. However, it is essential to note that regulations can be implemented without compromising the decentralized nature of digital assets. One of the most significant challenges facing regulators is the lack of compliance programs for digital assets. Compliance policies, customer identification, due diligence, and reporting are all critical components of a comprehensive compliance program. The absence of such programs makes it difficult for regulators to track and monitor digital currency transactions effectively.

While current regulations in the banking industry can be applied to digital currency transactions, it is crucial to develop regulatory policies specific to digital assets. Such policies should consider the unique characteristics of digital assets and ensure that they are effectively monitored and regulated. Additionally, law enforcement agencies must have the necessary tools and resources to enforce these regulations effectively. However, the lack of regulations surrounding digital assets poses a significant threat to the global financial sector. While enacting regulations may be challenging, it is essential to develop a regulatory framework that ensures the safe and secure use of digital assets. Such regulations should be comprehensive and consider the unique characteristics of digital assets, while also providing law enforcement agencies with the necessary tools and resources to enforce them effectively.

Digital currency scams affect economies in a variety of ways. The rise in the digital currency market globally attracts many participants. Therefore, enforcement of wrongdoing in the absence of regulations to regulate virtual assets is snowballing. The lack of regulations leads to countless problems such as market manipulation and insider trading (Khan and Hakami, 2021). Digital currency has been sensationalized globally and the properties of digital currency are attractive to users. Clearly, there is misuse and exploitation of digital assets that can cause a steep decline in their value of digital assets. Therefore, from a user perspective, protecting the digital asset from exploitation provides security and this is achievable through transparency of the data transactions. Both users and law enforcement benefit from regulations that can track transaction validity. Turner and Irwin (2018) postulate that cyber-attacks have been on the rise due to the acceptance of digital currency payment for these crimes.

Furthermore, digital currency scams can also lead to a loss of trust in the market, which can have a negative impact on the economy. If investors or users lose confidence in the integrity of the digital currency market, they may withdraw their investments or refuse to use digital currencies, resulting in a decrease in market value and overall economic impact. In addition, digital currency scams can also lead to financial losses for individuals and businesses. Scammers may use various tactics to steal digital currency, such as phishing scams, hacking, and Ponzi schemes. Victims of these scams may suffer significant financial losses, which can affect their finances and businesses, ultimately impacting the overall economy. The lack of regulations and oversight in the digital currency market can have far-reaching economic implications. Therefore, regulators and law enforcement agencies need to work together to establish regulations and enforce them to protect both users and the overall economy.

The Financial Industry Regulatory Authority (FINRA) is responsible for the registration of securities. Digital assets are not registered with FINRA. The lack of regulatory requirements prevents FINRA from tracking these assets. Due to the high demand for digital currency and assets, the government needs to take immediate action to implement laws and regulations to protect its citizens. Ivanovski and Hailemariam (2023) trust that establishing regulations is directly correlated with market satisfaction. Establishing a coherent global regulatory approach to regulate digital assets and currency will ensure investor protection and create a better framework for taxation and effective reporting systems (Hughs and Brown, 2018). Government action combined with lawmakers' efforts needs to establish clear guidelines to regulate the digital asset market to build moral utility.

Case Study: Intervention of the Australian Government to Regulate Digital Assets

Australian Transaction Reports and Analysis Centre (AUSTRAC) an Australian regulatory body imposes regulations on digital currency mandating the registration of all digital currencies on the exchanges (Hughs and Brown, 2018). Requiring the registration of digital currency ensures the monitoring and reporting of real-time data for law enforcement initiatives and creates transparency in the market for digital assets. Consequently, the Australian government has also implemented regulations on digital currencies by restructuring the process. Anyone who engages in digital currency must have a license issued by the Australian government (Australian Security and Investment Commission, 2022). Similarly, other jurisdictions can use the applications currently in force in Australia and implement regulations surrounding real-time monitoring of digital assets to create visibility and transparency around digital asset transactions. The blockchain technology that already exists at the base of digital currencies is a decentralized ledger for transactions where participants confirm transactions without the intervention of a central certifying authority utilizing either a private or a public network. Blockchain application has a robust platform that can manage a multitude of records pertaining to certificates of origin, bills of lading, invoices, import-export transactions, and paperwork (Duggan, 2022; Kramer, 2019; Susilowardhani et al., 2022). To empower law enforcement, all digital asset transactions require real-time monitoring to prevent and stop crimes (Prasad, 2022). The emergence and exponential growth of digital assets are creating threats for traditional banking institutions and shaking the global finance industry.

Digital assets have digital signatures that are controlled by mathematical algorithms that are used to validate a transaction's integrity and authenticity (Tanwar et al., 2021). The purpose of the digital signature is to act as a virtual fingerprint that is a unique identifier for the user and acts as a safeguard to the information contained in the digital document. Digital currencies use digital signatures to verify user access. Therefore, the users of digital currencies do not reveal the network's private key (Toradmalle et al., 2019). This safety feature for the users would continue as supported by the technology; however, each transaction that liquidates or funds can be accessed by law enforcement agencies to monitor activities of financial fraud.

The composition of the digital currency platform employs two facets: exchanges and wallets. The premise of both the exchanges and wallets is similar; however, users use exchanges to trade fiat currencies with each other and manage funds (Hoffman, 2018). An exchange facilitates the action to exchange currency for fiat and vice versa. Whereas a digital wallet uses wallet software that stores the fiat currency. There are opportunities for lawmakers to create regulations to mandate access to the exchanges and demand that the exchanges trade under the same laws as other financial instruments. The misuse of digital assets has supported several illicit activities. The core of blockchain technology where digital assets live provides anonymity. While anonymity is not immoral, protecting the economy from bad actors who use the strength of blockchain technology to hide their corrupt actions is immoral. The moral dilemma is that the same feature that protects the user from government control and provides autonomy also fuels illegal activities. The genesis of digital currency is to avoid centralization and provide a fiat alternative to government-controlled currency. Although the premise of digital currency creates equity, its execution, and management also occur in a secretive vacuum (Dierksmeier and Seele, 2016). Governments do not need to centralize the technology, the technology continues to function on its core principles; however, enacting regulations that provide law enforcement agencies and regulatory authorities with visibility and traceability of transaction data is desperately needed today.

2022 Internet Crime Complaint Center reported digital fraud losses of $246,212,432 as compared to the total sum of all cybercrimes reported at $463,424,524 (Jung et al., 2022). The data substantiates that digital fraud losses comprise 50% of total cybercrime losses. Furthermore, it is estimated that by 2025, financial crimes would reach $10 trillion. Digital crimes are on the rise and a significant obstacle to law enforcement investigation is privacy which creates the inability to track transactions.

The Internet Crime Complaint Center (IC3) is a division of the Federal Bureau of Investigation that monitors and enforces suspected internet-facilitated criminal activity. According to a communication announcement on May 4, 2022, from the IC3, Business Email Compromise (BEC) cost businesses and individuals $43 billion. In the digital asset world, the IC3 received an increased number of BEC complaints regarding the use of digital currency. The IC3 tracked down two leads, both of which were a direct transfer to a cryptocurrency exchange (CE) and a "second hop" transfer to the CE. In both situations, the IC3 found that the victims of the crime were unaware that funds were being sent to CE and then converted to cryptocurrency. In the direct transfer to the CE, the bad actor sends alerted wire information to the BEC victim, who then sends payment to the cryptocurrency custodial account on the CE. In a second hop transfer, the bad actor sends altered wire instructions to the BEC victim who then sends payment to the fraud victim (such as extortion, romance scams, etc.), the fraud victim sends funds to the bad actor and the bad actor creates a cryptocurrency account in the victim's name and then cashes out the cryptocurrency account. According to the 2018 Internet Crime Report, yearly and aggregate data for complaints and losses from 2014–2018, IC3 received a total of 1,509,676 complaints that resulted in $7.45 billion in total losses. The statistical data year over year shows a consistent growth that is detrimental to organizations and individuals.

As financial crimes keep rising, it is of utmost importance for lawmakers to design policies to protect the consumer. Money laundering involves activities of financial crimes that conceal the unscrupulous actor's identity and eventually provide access to funds that can be used for any purpose with minimal risks to the user. One infamous example of money laundering involves Liberty Reserve, a firm that was convicted for operating an unlicensed money-transmitting business and ultimately revealed over $6 billion in illegal operations (Dyntu and Dykyi, 2018). The Silk Road is another organization that operated as a part of the dark web allowing users to buy and sell products with each other anonymously. The platform conducted all transactions through Bitcoin to protect the participant's identities. Silk Road facilitated illegal products and illegal drugs on its platform. Platforms provided by organizations such as Liberty Reserve and Silk Road existed because of loopholes and unregulated freedom.

Digital asset exploitation leads to one of the most concerning activities: terrorist financing. Wang and Zhu (2021) reported that transfers of bitcoins to jihadis have even been supported by some sympathizers who have gone to lengths to publish books to teach how to transfer bitcoins from Western Europe and North America to Jihadists. Furthermore, it is reported that terrorist attacks in Jakarta in 2016, used Bitcoin to make virtual payments to finance terrorist activities. Terrorists have been enabled to use digital currency as a mainstream legal transfer tender within the financial system. Barone and Masciandaro (2019) reported that a study of 4,681 Ethereum networks (a digital asset exchange platform) found 2,179 accounts to be illegal. This was largely attributed to Distributed Autonomous Organization (DAO) failure. DAO is an emerging legal structure that has no central governing body where all users share a common goal to act in the best interest of the entity. DAOs are used for decision-making in a bottom-up management approach. DOA failure allowed hackers to divert $59 million from the $168 invested in Ehters (Barone and Masciandaro, 2019).

Financial Crimes Economic Impact

Financial crimes that are directly related to digital currency theft involve hacking and ransomware. Cybercriminals have detrimental effects on national security and financial systems. In 2017 a group of North Korean hackers used vulnerabilities in WindowsXP to launch a cyberattack through a ransomware attack named WannaCry 2.0. This attack affected various global participants, including Germany's railways, British National Health Service, and Boeing (Alfieri, 2022). Furthermore, North Korean hackers diverted $75 million in cryptocurrency from an organization in Slovenia and $24.9 million in funds from an Indonesian organization. It is estimated that North Korea pilfered approximately $1.75 billion in digital currency through these cyberattacks and used the funds obtained from these cybercrimes to finance nuclear capabilities of mass destruction. This is a very blatant example of how the theft of digital currency by cybercriminals can affect the global economy adversely. To protect their economies, every nation must enact laws to protect its organizations and individuals from cybercrimes that divert money through theft into financing terrorist activities that in turn cause an imminent threat to those very nations.

Cybercrime is a growing threat to national security and financial systems, every nation must take steps to protect its citizens and organizations from these crimes. The use of digital currency in these crimes has added a new layer of complexity, as traditional financial regulations and enforcement mechanisms may not be sufficient to prevent or investigate these crimes. Governments can take a range of steps to protect against digital currency theft and other cybercrimes. These steps might include developing and enforcing strong cybersecurity laws and regulations, investing in cybersecurity training and education for individuals and organizations, and supporting the development of new technologies and tools that can help prevent and investigate cybercrime.

In addition, governments can work together to share information and coordinate efforts to combat cybercrime on a global scale. This might involve cooperation between law enforcement agencies, financial regulators, and other organizations in different countries to track down and prosecute cybercriminals and recover stolen assets. Ultimately, it will take a coordinated and sustained effort by governments, businesses, and individuals to combat the growing threat of cybercrime and digital currency theft. By working together, we can help protect our economies and prevent these crimes from financing activities that threaten our national security and the safety of our citizens.

In the US, the Federal Trade Commission (FTC) is an agency that aims to protect consumers and enforces federal laws to prevent fraud, deception, and unfair business practices. FTC cannot keep pace with the scams related to the rapid evolution and popularity of digital currency. As proposed by Katarzyna (2019) there are two facets of innovation: one facet offers significant advances for the current process that are problematic by promising more efficiency, inclusion, and safety; while the other facet is unintended consequences of possible dangers and obstacles that the technology innovation may not have accounting for. Digital asset technology is not different from this premise as it is a new innovative technology that aims to revolutionize an age-old problem of fiat that was historically backed by gold; however, through time, it has lost all collateral properties and is just assumed as a currency that is operating on the faith of the users. The ability of the government to print as much as it would like is a factor that is not controlled by a common person's decision. Furthermore, the value of the currency is dependent upon the actions of the government of that jurisdiction. Economies have suffered great demise and discomfort due to negative government actions that have led currency fluctuations to teeter-totter and create chaos in the world. Numerous historical events have informed mankind to be cautious and look for alternatives to resolve lost trust in a centrally controlled fiat. Therefore, the invention of a decentralized currency resolves one issue of control; however, shifting total control to the user opens doors for criminals to swindle unsuspecting consumers leaving law enforcement agencies startled about how to protect the consumer. The continued growth in cybercriminal activity involving digital assets shows that there is a severe problem with this unique approach, and both good and bad actors are drawn to it.

Blockchain technology that creates a safe and trusted environment for consumers is also capable of building and tracking an environment for the regulatory enforcement agencies for the protection of the consumers. Lawmakers need to evolve exponentially to rectify this critical problem because, at the heart of a socially moral economy, consumer protection is essential. National governments have been forced to react due to the increased crime surrounding digital currencies. Central banks in many nations have reacted to digital currency crimes and have taken formidable steps toward consumer protection (Pieroni, 2018). In 2014, the US government banned Americans from using Russian banks, oil and gas companies, and other organizations following the invasion of Crimea. Russia's attack on Ukraine yielded the same sanctions; this time SWIFT banned the use of clearance of financial transactions from and to Russia, creating a financial upheaval for Russia. It is estimated that Russia loses $50 billion due to western sanctions annually (Flitter et al., 2022). More government intervention is expected as lawmakers step up to boost the scrutiny of multiple digital assets. For example, the dollar is the reserve currency for many global payments, the US can use sanctions as a diplomatic tool. However, diplomatic sanctions only work with digital assets that are pegged by the US dollar. Since various cryptocurrencies exist, countries like Russia can avoid digital currencies that are pegged to the dollar to circumvent the US sanctions. Russia is developing its own proprietary digital currency that will not convert to U.S. dollars (Flitter et al., 2022).

Other nations such as North Korea and Iran have utilized digital currencies to avoid sanctions issued by the Western nations. While digital currency has been in existence for over 20 years, governments are slow to accept it as legal tender. Although digital currency provides safety to the investor, the prominence of cybercriminal activities and theft leaves the consumer unprotected. Government intervention is needed to earn investor trust. Policymakers need to consider structuring clear regulations for digital assets by imposing taxes on internet purchases thereby allowing government regulators access to the data for minimizing corrupt activity. Kamau (2022) believes that digital asset management requires a concrete legal framework to fully experience the advantages of the technology.

Although regulators are slow to move, policymakers are intervening to construct a legal structure to protect consumers. The Senate Agriculture Committee which is responsible for overseeing the Commodities Futures Trading Commission (CFTC) recently passed a bipartisan bill that grants the CFTC authority over digital currency transactions entering the commodities law (De, 2022). Digital Commodities Protection Act of 2022 requires digital commodity platforms to prohibit abusive trading practices, remove and disclose conflicts of interest, maintain sufficient financial resources, provide a strong cybersecurity program, protect consumer assets, and report suspicious transactions (Stabenow, 2022). From the genesis of Bitcoin in 2008, thousands of digital assets have been developed and now represent about a trillion dollars in wealth. It is estimated that one in five Americans have used digitally traded assets.

The lure of digital assets includes a robust design to make financial systems more accessible, the volatility in digital asset pricing, cybercriminal activities, and trading platform abuse have resulted in billions of dollars of losses. The industry leaders Bitcoin and Ether are commodities that are traded on trading platforms that are not federally regulated. Due to the lack of federal regulations, consumer protection lacks the transparency and fairness that is expected from the financial system. Lack of governance will continue to hinder consumer protection as vulnerability to fraud and manipulation rises. The Digital Commodities Protection Act of 2022 aimed to close regulatory gaps by requiring all digitally traded commodity platforms to register with the CFTC and uphold reporting of abusive trading practices, and most importantly recognize that financial agencies have a role in regulating digital assets that are not commodities. The Digital Commodities Consumer Protection Act of 2022 gives authority to CFTC to oversee all digital commodity transactions on exchanges by default as the law stipulates that digital securities must be registered through the CFTC.

On January 9, 2023, according to a press release from CFTC, Avraham Eisenberg, a 27-year-old U.S. citizen and self-described digital art dealer, who is alleged to be the mysterious hacker behind the $110 million manipulation of digital asset exchange run by Mango DAO. This case is the first case of Oracle manipulation, which is the manipulation of an Oracle smart contract that is compromised by an attacker that leads to system failure, and theft. In the press release, CFTC goes on to say that Eisenberg unlawfully misappropriated over $110 million in digital assets from Mango Markets, a decentralized digital asset exchange through oracle manipulation. Eisenberg created anonymous accounts and established large leveraged positions artificially spiking the price by 13-fold during a 30-minute span and then cashed out his illicit profits by using the artificially inflated price of his collateral to withdraw over $110 million in digital assets from Mango Markets catastrophically draining the platforms assets that belonged to other users (Murphy et al., 2023). Additional government agencies are clearly defining digital assets. Corbet et al. (2018) assert that the IRS treats cryptocurrencies as property and not currency. Therefore, the IRS requires Form 8949 to report capital gain or loss on the sale of Bitcoin. Coinbase, a popular digital currency exchange, was directed by a court in 2017 to provide the IRS with taxpayer addresses, dates of birth, name identification, and records of all transactions. Furthermore, IRS penalizes unreported revenues from gains of digital currency sale and civil penalties of 5% of the unpaid tax amount per month capped at 25%. The creation of well-defined, clear rules, which classify digital assets as real property, and then designing penalties and civil assessments is a good regulatory initiative.

Similarly, the SEC has jurisdiction over the financial services industry. Digital assets are like capital market products and, by default, fall under the jurisdiction of the SEC. Moreover, The U.S. Department of the Treasury's Financial Crimes Enforcement Network (FinCEN) is a government agency that monitors networks to punish companies and individuals engaged in financial crimes reported in a January 18, 2023, press release identifying virtual currency exchange Bitzlato as a primary money laundering concern in connection with Russian illicit finance (FINCEN, 2023). FinCEN identified that Bitzlato, which operates outside of the U.S. plays a critical role in laundering Convertible Virtual Currency (CVC) by facilitating illicit transactions for ransomware users operating in Russia. CVC is a digital asset that has equal value to real currency and acts as a substitute for real currency. The U.S. remains committed to identifying networks such as Bitzlato and disrupting such networks from gaining access to the U.S. financial system.

The regulation of virtual currencies is an important issue that requires collaboration between regulatory agencies such as FinCEN and the Bank Secrecy Act (BSA) to ensure compliance with the provisions of the BSA. Cryptocurrency poses a significant risk to the global financial system due to its potential use for criminal activities such as terrorism financing, money laundering, and cybercrimes. Therefore, lawmakers need to regulate digital assets in a standardized fashion to prevent criminal activities and protect the world economy.

One of the key challenges facing the regulation of virtual currencies is the anonymity of transactions enabled by blockchain technology. Criminals often rely on the fact that their transactions cannot be traced or reprimanded for engaging in illegal activities. To address this challenge, lawmakers must focus on removing anonymity from the equation by implementing robust identification and verification procedures for virtual currency users. Moreover, technology can play a significant role in enabling the traceability and validity of digital transactions. Lawmakers can leverage the strengths of blockchain technology to program tracking mechanisms that run algorithms and transaction details to help ensure the traceability and validity of digital transactions.

Regulating virtual currencies is critical to prevent criminal activities and protect the global financial system. Lawmakers need to collaborate with regulatory agencies such as FinCEN and BSA to ensure compliance with the provisions of the BSA. Furthermore, removing anonymity from the equation and leveraging technology to enable the traceability and validity of digital transactions can help address the challenges facing the regulation of virtual currencies.

Virtual currency networks and exchanges have been subject to the BSA money transmission requirements. BSA regulations apply to individuals authorizing, securing, transmitting, accepting, exchanging, and distributing virtual currencies (Albrecht et al., 2019). FinCEN does not regulate virtual currency users, but regulates exchanges and administrators, specifically money transmitters. BSA on the other hand oversees cryptocurrency regulations in the U.S. via the FinCEN treasury alliance (Toscher and Stein, 2018). There is an opportunity for both agencies, FinCEN and BSA, to collaborate to ensure third-party administrators and exchanges for digital assets adhere to provisions set out under the BSA. Lawmakers across the globe must agree to regulate digital assets in a standardized fashion to minimize and prevent criminal activities. Cryptocurrency poses a risk to the global financial sectors due to threats of terrorism and criminal activity financing. The technology is a target of organized crime due to transaction anonymity and lack of regulation. Money laundering and cybercrimes are also rampant and pose a threat to the world economy. In addition, lawmakers must focus on the anonymity of the technology. Criminals are running rampant because blockchain technology offers anonymity and people who engage in illegal activities are relying upon the fact that they will remain anonymous and therefore cannot be traced or reprimanded. If anonymity is removed from the equation, criminally transpired transactions will cease due to fear of being caught. Another factor to consider is to play to the strengths of the technology itself and program tracking mechanism that alters authorities by running algorithms and transaction details to help ensure the traceability and validity of the digital transaction.

Structure

Cryptocurrency transactions have created ease of utility in the past decade. The main utility of cryptocurrency is the purchase and sale of goods and services without relying on financial institutions. The most popular example of cryptocurrency is Bitcoin, created by Satoshi Nakamoto in 2008 and since its discovery, many more cryptocurrencies have propped up in the market. It is believed that digital assets have many recurrent arbitrage opportunities and there is a lot of unknown about these assets (Akyildirim et al., 2020; Corbet et al., 2018). The transformation of the current technological revolution is creating new business models through the innovation of information technology represented by the Internet of Things, big data, cloud computing, artificial intelligence, and blockchain.

Indeed, cryptocurrency transactions have revolutionized the way people conduct financial transactions and have created a decentralized system that eliminates the need for intermediaries such as banks. The use of blockchain technology, which is a decentralized and distributed digital ledger, ensures that transactions are transparent, secure, and immutable. Apart from the ease of use, cryptocurrencies have also presented a new investment opportunity, with many investors buying and holding cryptocurrencies as a form of investment. However, the volatility of the cryptocurrency market and the lack of regulation in some jurisdictions make it a risky investment.

Furthermore, the integration of blockchain technology with other emerging technologies such as the Internet of Things, big data, cloud computing, and artificial intelligence is creating new business models and opportunities. For instance, blockchain-based smart contracts enable the execution of contracts in a transparent and automated manner, while the use of big data and artificial intelligence in cryptocurrency trading can help to identify arbitrage opportunities and optimize trading strategies.

However, as with any new technology, there are still many unknowns and challenges associated with cryptocurrencies and blockchain. These include issues related to scalability, interoperability, and regulation, which need to be addressed for the full potential of these technologies to be realized. Thus, cryptocurrency transactions and blockchain technology have presented a paradigm shift in the way financial transactions are conducted, creating new opportunities for innovation and growth in various sectors of the economy.

Trust is the belief between people transacting with each other that premises around positive outcomes and enables individuals to simplify many decision-making endeavors. Today in an actively connected world, many individuals perform the bulk of their transactions in business, entertainment, news, and critical healthcare services in an online setting. Trust plays a critical role in online transactions because we lean towards relying on heuristic trust to simplify complex decision-making. A modern overloaded and technology-saturated society driven by information and communication technologies must provide simplification for the users. Therefore, trust must be at the core of the next generation of technology. Safeguarding personal information is more crucial than ever. The increase in consumption of data is on the rise. According to Vailshery (2022), the number of IoT devices is expected to triple from 9.7 billion devices connected to the internet to an estimated 29 billion connected devices by 2030. Furthermore, it is estimated that an average American spends more than 1300 hours on social media of all types.

According to the World Economic Forum, global internet users spend an average of 2 hours and 27 minutes per day on social media platforms (Buchholz, 2022). Hyperconnectivity over technology is clearly on the rise. As connectivity continues to grow and accelerate, organizations are finding it increasingly difficult to accurately provide secure platforms for data storage that provides robust data privacy and security to earn and maintain consumer trust. Sherchan (2013) proposed a trust management system for application in online social networks and defined three key dimensions: trust information collection, trust evaluation, and trust dissemination. The trust information collection process requires information gathering and organization into a user-friendly format and ensures the computation of trust. Many organizations fail to comprehend this critical data input needed on the network. Trust evaluation comprises historical transactional data between users and an online system. Trust evaluation is the use of data to compute trust values between nodes or the network. Trust dissemination makes use of computed trust values with specificity, such as targeted suggestions to users such as a movie recommendation to a user based upon trust in another similar user (Solodan,

2019). Why is the impact of trust so important? Understanding trust helps improve network performance by rewarding trusted nodes and removing untrusted nodes to support real-world human-to-human trust relationships.

While trust management systems have been employed for online reputation systems such as eBay, where buyers rate sellers on the quality of the product and the transaction. Blockchain technology facilitates transactional validity and transparency to increase trust among participants. The decentralized system provides interoperable, interconnected, and compatible with anonymous contextual chronological and resilient properties to create an interface for open data to users. Blockchain can create a consolidation of data that currently exists on many online services but the inclusion of several online networks and social media platforms presents a challenge to trust. In the saturated world of social media platforms and other online networks, blockchain can benefit from data fusion across these networks and create trust.

Framework

We hope our framework will provide industry leaders seeking clarification and understanding of blockchain with detailed and easy-to-understand applications of blockchain technology. As illustrated in Fig. 3, there are three layers to recognize.

1. Physical layer
2. Current layer
3. Blockchain layer

Physical Layer

The physical layer is the current state of the finance industry where different participants actively engage in business transactions. There are two sides to these business transactions: demand and supply. Demand-side constraints are volatility in income, geographical barriers, information asymmetry, literacy, and lack of trust. In the global market for financial services currently, the demand for services is very volatile due to income disparity. The products sold in the financial services industry are focused on a fee-based model and do not support the investment and financial goals of most individuals. Only affluent clients can afford to pay for the service. In addition to income volatility, geographical barriers play an important role not just in investment, but in money flow. There are many world citizens who currently do not have access to brick-and-mortar banks to serve the needs of cash flows; therefore, consumers in developing countries must rely upon many hours of transportation to access cash for their daily needs, a term called unbanked. Apart from the unbanked of the world, globalization creates an immediate and fast need for money to flow across boundaries. Presently intermediaries that facilitate the flow of money across borders severely rely upon slow legacy systems that cost the consumer high

Provides
- Conducive legal and regulatory frameworks
- Enables digital infrastructures
- Supports ancillary government systems

Blockchain Layer

Cost, speed, transparency, security and immutability

Current Information Layer   Lack of digital infrastructure

Lacks legal and regulatory frameworks   Lack of ancillary government support systems

Supply
- High operating costs
- Inefficient legacy systems
- Limited competition and innovation

Physical Layer Constraints

Demand
- Volatility in income
- Geographical barriers
- Information asymmetry
- Literacy and lack of trust

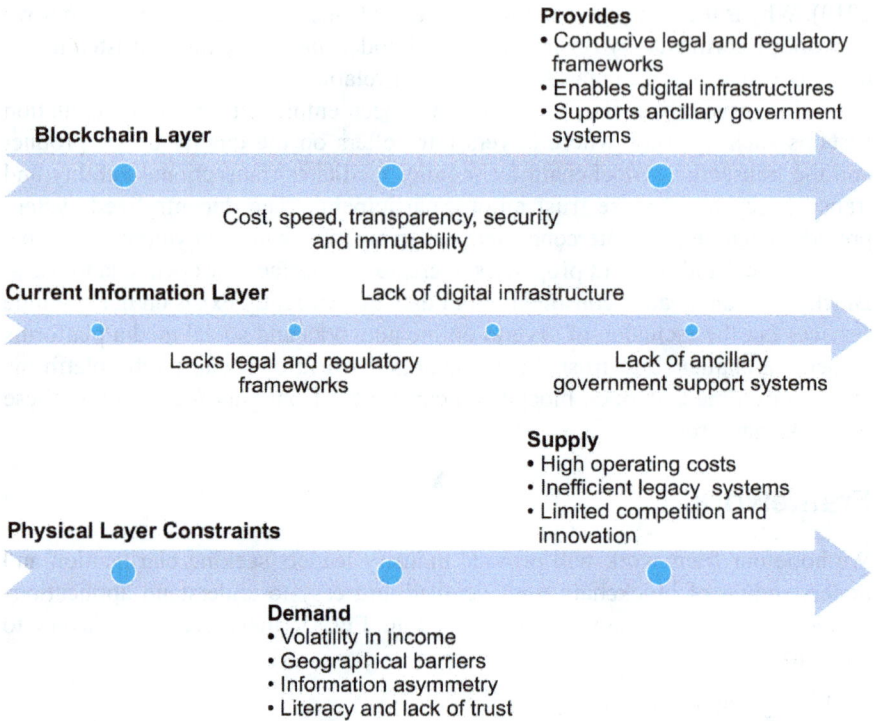

Blockchain application in the financial services industry.

Figure 3: Framework financial services industry (Source: Author's own work)

transaction costs and time inefficiency that can take a transaction a few days to process. In addition, tracking of a transaction is spotty at best, and due to many manual approvals and overrides it is not easy to get the correct information to trace transactions easily. Legacy systems and heavy human interaction create information asymmetry and causes trust erosion.

On the supply side, the physical constraints are high operating costs, inefficient legacy systems, and limited competition and innovation. Organizations compete for resources and reducing operating costs is a massive strategic advantage. For instance, currently, financial institutions are struggling with massive amounts of paperwork that must be collected on each account to establish the identity of the client known as Know Your Client (KYC), this is a regulatory requirement. The collection and input of data are done manually. Sometimes, the information collected is either incorrect or entered manually into the system due to human error. Removing human obstacles and depending upon an automated real-time environment where clients can enter their own information, not only creates a digitally verifiable identity for the client, it also ensures that slow-moving and manual processes are removed to gain efficiencies.

Current Layer

Shared information compatibility is perhaps the primary reason to apply blockchain technology to the finance industry. The current technology lacks the important infrastructure to support the regulatory framework. The finance industry works in a centralized format where firms operate within a very strict framework that is fragmented. Access to data is extremely scarce to the employees of the firm and only employees that have access to the highest security are privy to most data. This practice is common in the industry. However, by centralizing and removing access to data, most employees must rely upon other departments to produce paperwork, portfolios, and special analysis as needed which slows down the process of information flows and eventually creates time and cost inefficiencies. The advent of blockchain technology has come a long way since the development of Bitcoin; however, lawmakers have not been able to update laws to prevent maleficence. Adopting blockchain technology grants access to regulatory and compliance agencies, where agencies can see the data and transactions in real-time to monitor for inconsistencies. This feature alone leapfrogs the current infrastructure through the next few cycles of innovation and creates tremendous value for organizations.

Blockchain Layer

At present, financial institutions share information in fragmented, slow, and expensive legacy systems. The new framework offers professionalism, technology, effectiveness, service, and scale, all delivered instantaneously. Professionalism is achieved through a team of geniuses that take advantage of the smart contract feature on the blockchain. Smart contracts feed into an advanced technological feature that facilitates the elevated use of technology. The decentralized feature of the blockchain creates high efficiency and effectiveness. Transparency is achieved through high-quality service and traceable and tamper-proof properties of the blockchain creating scalable real-time security. The use of this framework creates a blockchain mode where all businesses are automatically executed per the contract that is layered on smart contract technology. Smart contracts utilize operation logic and execution conditions to process transactions. Logic is built into the core of financial transactions; therefore, automated execution can be affected using smart contracts. In the application layer of blockchain, technical programming can create elevated financial information fusion and sharing services that can better integrate with regulators and compliance initiatives.

Application

Wide-ranging technological evolution that contains the Internet of Things, big data, cloud computing, artificial intelligence, and blockchain is at the center of driving information toward digitization and building new ecosystems. Intelligently building information technology is the fabric of tomorrow's information technology platform. The financial field has heavily adopted emerging cutting-edge technologies to pave the way for major innovative modifications in the financial markets and financial service industry dubbed fintech. At the heart of fintech technology smart contracts, smart cities, and smart buildings are all emergent technologies. The global evolution of technology is currently in a transformative stage that continues to deepen. Largely due to preexisting issues that came to light in the current pandemic of COVID-19 has forced industry players to seek new technologies. The outbreak of COVID-19 fueled the advent of the digital economy which is quickly rising to the challenge and is the new driving force for promoting economic and social development. The phenomenon created by the pandemic disrupted businesses and individuals worldwide. Basic factors that play an important role in our lives such as work, food, manufacturing, shipping, and severe shortfalls in supply chains exacerbated functions of our daily life and created immense panic. Organizations rely upon emerging technologies to break time and space constraints. For instance, building and linking digitally intelligent databases to create a fully interactive digital economy creates a new ecosystem that fully collects, shares, and makes use of data in an orderly and trusted fashion to sustain and develop the economy and society. The traditional mode of business operations is screaming to achieve an upgrade where lifestyle, social governance, and industries will be systematically transformed into digitization and networking.

Currently, IoT applications used for digitization rely on centrally adopted technology and operation modes. Due to the centralized nature of the technology, there are several concerns regarding trust, data value, and business models. In terms of technological compatibility, the shortcomings include poor architecture and a slow interface between equipment and the system. Due to the poor infrastructure of conventional technology, there is a loss of efficiency, higher costs, and weak scalability. Furthermore, data privacy and security are poor, and ownership of data is imprecise. IoT applications run on the principle of data values; however, data values generated by the IoT are unable to apprehend effective transmission of data value to make full sense of the data. There is currently a huge gap that exists between conventional and emerging technologies to support industry business models to create a suitable blueprint for operations and maintenance of data. Introducing blockchain as a solution provides a well-rounded opportunity for advancing industry business models and recreating data efficiency, security, and storage by building a decentralized IoT platform that promotes innovation. The integration of blockchain technology deepens IoT technology by decentralizing, creating a trusted and distributed ledger, reducing the cost of system interconnection, and improving open data sharing by providing user privacy and system security. These characteristics of blockchain technology are the building blocks of the next generation of technology. Every bit of data will live on the chain in the future! It is wise for organizations to seriously investigate blockchain technology to transition smoothly into the future of business.

A comprehensive literature review revealed key technological aspirations currently in force. In China, mobile phone and tablet technology are being used to relate data through iBeacon ID which houses users' personal data to provide directions (Yang et al., 2019). In Italy, wearable devices, processing centers, and multimedia walls are utilizing user location to provide wide-ranging content information for museum artwork (Alletto et al., 2016). Intelligent mobile robots in Korea are constructed on Wi-Fi signal operability (Lee et al., 2017a). In the US, bar codes, tags, and RFID are being used to track and locate assets (Asset Management and Tracking, 2023). In Singapore, intelligent plugs, multipurpose nodes, BLE sensors, and intelligent gateways are being used to control intelligent grid systems for energy control (Viswanath et al., 2016). In the Netherlands, sensors and cloud-based intelligently facilitate the management of groups of people to measure CO_2, light intensity, and humidity (iSCOOP, 2017).

Australia is integrating a big data platform for indoor environment analysis to measure oxygen levels, luminosity, and hazardous gasses through Hadoop—a big data ingestion and analytic technology (Bashir et al., 2016). In Canada, body sensor readings and personal health are monitored by systems (Li et al., 2011). All the technologies suggest the new wave of technology needs. The integration of IoT and blockchain technology creates advantages for data storage. Blockchain uses two functions: block header and block body. The block body stores all the verified transactions, and the block headers store specific metadata such as the hash of the previous block, the hash of the current block, and timestamps. Therefore, blockchain technology is effectively applicable in almost all domains of IoT. Wu et al. (2020) research demonstrated the use of IoT and blockchain technology in pledging movable assets in financial services.

Moveable assets are properties owned such as works of art that one can take with them. Tracking movable property is challenging because financial institutions cannot access real-time information on the moveable pledge asset due to insufficient access management and information sharing surrounding the movable property assets among all stakeholders. Traditionally there is no current system to monitor or track movable property to validate the asset for securing loans. Furthermore, some bad actors purposefully pledge movable assets and take fraudulent loans using loopholes due to the lack of information symmetry that creates liabilities and losses for financial institutions. The union of IoT and blockchain technologies provides a robust solution for tracking, monitoring, and tracing the moveable asset. The transaction data stored on the blockchain provides security and transparency to the lending institution thereby eliminating duplicate pledges and reducing fraudulent activity creating cost savings and reputational trust. There are several steps involved in describing the transactions between a piece of art and a loan:

1. **Loan Application:** A loan application is processed to secure a loan against the art piece. The loan application is submitted to the financial institution to secure the loan.

2. **Authorization:** The bank reviews the loan application and receives data on the art piece to approve the loan.
3. **Data Collection:** The borrower and the lender share data on the art piece and monitoring data is generated.
4. **Monitoring Data:** The transactions recorded on the blockchain are then validated and available to all users on the platform, which is an important evidentiary component for judging the repayment of the loan.
5. **Loan Approval:** After all data is reviewed and asset status is validated, the financial institution makes the final decision to approve the loan for the art piece, and the loan is funded.

Case Study: Christie's Auction House

Christie's and Sotheby's are long-standing landmarks in the multi-billion-dollar global art auction industry. Auction houses have evolved over the centuries and pivoted in business strategy to continue to grow their popularity. Since their establishment in the late 1700s, auction houses have progressed radically, especially in the face of adversity and business progression. Auction houses have introduced innovations such as black-tie events in the 1950s to evolve and connect auction houses' reach to an affluent worldwide audience. Therefore, auction houses are no strangers to a pivoting business model. The COVID-19 pandemic crisis had a severe impact on auction houses. And just like that auction houses had to respond to severe losses and change. Christie's, Phillips, and Sotheby's together observed a 79% decline in revenue during the second quarter of 2020 compared to the same period in 2019 (Loss, 2020). Additionally, the Financial Times reported that in 2020 Christie's sales plummeted from $2.3 billion in revenue to $364 million as compared to 2019. To combat declining sales, and the new uncertainty of social distancing that prevented auction house patrons from attending extravagant in-person galas, the auction houses had to recreate their business model once again.

Christie's quick response to Covid-19 created a worldwide live auction that was conducted in succession in Hong Kong, Paris, London, and New York and produced $420,941,024 in revenue supporting a sell-through rate of 94% (Christie's Annual Report, 2019). Christie's reported that more than 80,000 participants tuned in to experience the new spectacle and 60,000 watched through social media in Asia. In 2021, Christie's reported that 3.3 million consumers participated in live-streamed auctions through various online sources (Christie's Annual Report, 2021).

Christie's upholds that continued leadership innovation in digital technologies has yielded great enthusiasm for collectors and reaffirmed its auction market position. The immense popularity of online auctions created Non-Fungible Tokens (NFTs). NFT is a blockchain-based token that represents real assets such as a piece of art or digital content. For the first time, Christie's experienced a 30% growth of new customers in 2021 with the addition of millennials (Christie's Annual

Report, 2021). Christie's pioneered a remarkable breakthrough through NFTs to attract a younger client base. Christie's reported that they sold 150 million dollars in NFT. The head of digital art and online sales, Noah Davis, says that NFTs are here to stay. Blockchain-driven initiatives are prominent among market leaders and iconic brands are swiftly seeking NFTs in their marketing efforts.

In May of 2017, blockchain technology gained a major win with the announcement of R3 had raised US$107 million through the collaboration of investors like Intel, Bank of America Merrill Lynch, UBS, HSBC, and the Singapore government to develop blockchain technology to be used in major banks (Shieber, 2017). The success of R3 positioned the firm at the confluence of technology and finance. The global financial industry's success rests on the heels of technological inventions. Technology is a major component in creating a strategic advantage for industry leaders. The major players in the global finance field understand the critical importance of revolutionary technology and are willing to take the risk to adopt early technologies such as blockchain to gain efficiency. Historically the banking industry had to face staunch competition, increased cost pressure, and increased regulations that have damped their profit margins. The traditional banking system housed all functions of banking in-house. However, the high cost and regulatory pressure kept increasing and banks finally had to falter to use third parties to create more checks and balances which increased the cost of doing business substantially. As a result of the R3 prototype conjecture, Bank of America Merrill Lynch has poised itself to be an industry leader. Many industry insiders share the massive leapfrogging Bank of America Merrill Lynch has achieved in technological innovation through the adoption of a blockchain-based infrastructure.

Blockchain technology is a real-time technology that allows encrypted data on any type of transaction to be shared among all participants (companies, people, and government). Bitcoin created the hype for blockchain technology. The misconception that bitcoin and blockchain are synonymous with each other created more of a smoke-and-mirror scenario for organizations. The mystery around blockchain technology remained synonymous with bitcoin regardless of the technology's obvious advantages of application in various industries. Blockchain infrastructure is a part of Web 3.0 which is the next generation of the internet to manage digital information in a distributed ledger format giving users the ability to create value and authenticate digital information. In the financial sector, blockchain applications are already very successful in banking functions including international cross-border payments, capital market, and trade finance, regulatory compliance and audit, and defense against money laundering. The core of blockchain technology creates a real-time (continually reconciled) database that uses a decentralized digital ledger to secure, validate, and trace transactions. It is estimated that the majority of the biggest investment banks are implementing blockchain technology and are estimated to save up to 30% in cost savings resulting in an estimated $8 billion to 12 billion in savings (Yang et al., 2019). Many clearing houses still rely upon fragmented systems and legacy systems that were created due to regulatory checks and balances and did not communicate with each other Therefore, the clearing houses must rely upon a very manual reconciliation process between different members of the intuition.

The hunger for new innovative technologies is at the core of traditional banking transformation to newly developed, automated, and efficient platforms to create value for all stakeholders. For instance, in the past, a bank's payment processing center has been an extremely manual process mainly considered a support function for processing payments rather than generating revenues. However, the application of blockchain-based technology creates a support function into a revenue-generating function by automating the process and removing cumbersome manual tasks that require human intervention. The finance industry has traditionally struggled with similar issues due to the nature of the complexity of the transactions in the field. A high level of privacy and access is practiced in the industry limiting critical access to transaction data with the network. However, the use of blockchain technology provides a large consortium of document and process management through automation, validated data security for shared data, easy integration, and algorithms that create a pluggable consensus that enables compliance and local regulation inclusion.

Conclusion

There is increasing hype for a digital currency that is persistently shifting and transcending national borders. Although digital currency is very widespread it has been afflicted by illegal activity that has followed suit forcing law enforcement to question its effectiveness. The government has intervened in several legal and regulatory matters regarding digital currency and fraud, and there is a substantial ever-expanding gap between digital currency and organized crime. The application of the law has been slow-moving and therefore created challenges. Digital currencies' main characteristics are decentralization and anonymity. Criminal activities are pervasive and harmful to all industry participants. Activities such as hacking, ransomware attacks, money laundering, counterfeiting, and drug trafficking are some illegal activities that have adopted cryptocurrencies to circumvent the law. Law enforcement faces a precipitous challenge when digital assets are used by criminals to purchase drugs, weapons, and other contraband anonymously.

The impact of financial crimes can be devastating to the world economy. The blockchain technology that creates a safe and trusted environment for consumers is also capable of building and tracking an environment for the regulatory enforcement agencies for the protection of the consumers. Lawmakers need to evolve exponentially to rectify this critical problem because, at the heart of a socially moral economy, consumer protection is essential. National governments have been forced to react due to the increased crime surrounding digital currencies. Our framework for the financial industry focuses on three layers: the physical layer, the current layer, and the blockchain layer. The physical layer is the current state of the finance industry where different participants actively engage in business transactions. There are two sides to these business transactions: demand and supply. Demand-side constraints are volatility in income, geographical barriers, information asymmetry, literacy, and lack of trust. On the supply side, the physical constraints are high operating costs, inefficient legacy systems, and limited competition and innovation. Organizations are competing for resources and reducing operating costs is a massive strategic advantage. For instance, currently, financial institutions are struggling with massive amounts of paperwork that must be collected on each account to establish the identity of the client known as Know Your Client (KYC), this is a regulatory requirement.

Shared information compatibility is perhaps the primary reason to apply blockchain technology to the finance industry. The current technology is lacking the important infrastructure to support the regulatory framework. The finance industry works in a centralized format where firms operate within a very strict framework that is fragmented. Access to data is extremely scarce to the employees of the firm and only employees that have access to the highest security are privy to most data. This practice is common in the industry. However, by centralizing and removing access to data, most employees must rely upon other departments to produce paperwork, portfolios, and special analysis as needed which slows down the process of information flows and eventually creates time and cost inefficiencies.

At present, financial institutions share information in fragmented, slow, and expensive legacy systems. The new framework offers professionalism, technology, effectiveness, service, and scale all to be delivered instantaneously. Professionalism is achieved through a team of geniuses that take advantage of the smart contract feature on the blockchain. Smart contracts feed into an advanced technological feature that facilitates the elevated use of technology. The decentralized feature of the blockchain creates high efficiency and effectiveness. Transparency is achieved through high-quality service and traceable and tamper-proof properties of the blockchain creating real-time security that is scalable. The use of this framework creates a blockchain mode where all businesses are automatically executed per the contract that is layered on smart contract technology.

The appetite for new innovative technologies is at the core of traditional banking transformation to newly developed, automated, and efficient platforms to create value for all stakeholders. For instance, in the past, a bank's payment processing center has been an extremely manual process mainly considered a support function for processing payments rather than generating revenues. However, the application of blockchain-based technology creates a support function into a revenue-generating function by automating the process and removing cumbersome manual tasks that require human intervention.

Chapter Summary

The finance industry has been one-sided since inventing blockchain technology for fiat and cryptocurrency. Although the advent of blockchain technology took place in 2008 with the introduction of Bitcoin, the technology is greatly misunderstood as a facilitator of cryptocurrency-based transactions only. While digital currency is very widespread and deeply associated with blockchain technology, the technology is useful in all aspects of the finance industry. This chapter focused on the lack of regulations surrounding digital assets that threaten the global financial sector. It delved into the challenges that lawmakers currently face to properly enforce regulation of digital assets and to explore exploitation that includes cybercriminal activities, money laundering, and terrorist financing are discussed and how blockchain can be an effective solution in eradicating this misuse. The impact of financial crimes on economies is outlined using a conceptual framework that is constructed for ease of use by the reader and exhibits how blockchain technology would theoretically and practically operate in the finance industry. The construct of the framework is divided into three sections: the physical layer, the current layer, and the blockchain layer for simplifying the integration of blockchain technology in the finance sector.

As financial crimes keep rising it is of utmost importance for lawmakers to design policies to protect the consumer. Money laundering involves activities of financial crimes that conceal the unscrupulous actor's identity and eventually provide access to funds that can be used for any purpose with minimal risks to the user. The composition of the digital currency platform employs two facets: exchanges and wallets. The premise of both the exchanges and wallets is similar; however, users use exchanges to trade fiat currencies with each other and manage funds (Hoffman, 2018). An exchange facilitates the action to exchange currency for fiat and vice versa. Whereas a digital wallet uses wallet software that stores the fiat currency. There are opportunities for lawmakers to create regulations to mandate access to the exchanges and demand that the exchanges trade under the same laws as other financial instruments. The misuse of digital assets has supported several illicit activities. The core of blockchain technology where digital assets live provides anonymity. While anonymity is not immoral, protecting the economy from bad actors who use the strength of blockchain technology to hide their corrupt actions is immoral. The moral dilemma is that the same feature that protects the user from government control and provides autonomy also fuels illegal activities. The genesis of digital currency is to avoid centralization and provide a fiat alternative to government-controlled currency.

Furthermore, Christie's case study depicts a remarkable breakthrough through NFTs to attract a younger client base and the creation of a new digital asset category of NFTs that are currently creating positive waves in the financial market. Blockchain technology is a real-time technology that allows encrypted data on any type of transaction to be shared among all participants (companies, people, and government). Bitcoin created the hype for blockchain technology.

The misconception that Bitcoin and blockchain are synonymous with each other created more of a smoke-and-mirror scenario for organizations. The mystery around blockchain technology remained synonymous with Bitcoin regardless of the technology's obvious advantages of application in various industries. Blockchain infrastructure is a part of Web 3.0 which is the next generation of the internet to manage digital information in a distributed ledger format giving users the ability to create value and authenticate digital information. In the financial sector, blockchain applications are already very successful in banking functions including international cross-border payments, capital market, and trade finance, regulatory compliance and audit, and defense against money laundering. Many clearing houses still rely upon fragmented systems and legacy systems that were created due to regulatory checks and balances and did not communicate with each other Therefore, the clearing houses must rely upon a very manual reconciliation process between different members of the intuition. The appetite for new innovative technologies is at the core of traditional banking transformation to newly developed, automated, and efficient platforms to create value for all stakeholders.

Take Away Lessons from the Chapter

1. The invention of blockchain technology is gravely misunderstood by the masses. The finance industry has sensationalized the association of blockchain technology with Bitcoin and most believe that blockchain is not a technology but a facilitator of Bitcoin and another cryptocurrency.
2. Digital currency is widespread and deeply associated with blockchain technology, however, the technology is useful in all aspects of the finance industry, not just fiat currency.
3. Blockchain technology is a real-time technology that allows encrypted data on any type of transaction to be shared among all participants (companies, people, and government). Bitcoin created the hype for blockchain technology.
4. Though regulators are slow to move, policymakers are intervening to construct a legal structure to protect consumers.
5. The misuse of digital assets has sustained several illicit activities that can be reinforced using anonymity, which is a core feature of blockchain technology.
6. Tracking movable property is challenging because financial institutions cannot access real-time information on the moveable pledge asset due to insufficient access management and information sharing surrounding the movable property assets among all stakeholders. Additionally, there is no current system to monitor or track movable property to validate the asset for securing loans.
7. Blockchain technology creates a safe and trusted environment for consumers in addition, also capable of building and tracking an environment for the regulatory enforcement agencies for the protection of the consumers.

Chapter Exercises

1. Conduct a brief class discussion about the current deficits in technology innovation in the finance industry. Next, create a blockchain-based solution to identify the benefits of blockchain to remove the identified deficiencies.
2. Assign teams to apply blockchain technology solutions to different sectors of the finance industry to showcase where and how blockchain can make a difference.
3. Break the class into teams and use the framework provided in this chapter to perform a SWOT analysis of the framework.

Blockchain Applications and Uses: Healthcare Industry

The potential exists for blockchain-based technology to significantly improve the healthcare industry. However, there are challenges posed by the widespread adoption of blockchain-based technology in health care. This chapter begins with a brief refresher of what blockchain technology is and then covers a variety of benefits and challenges to utilizing blockchain, specifically in health care. Among the purported benefits of blockchain is its potential to improve the accuracy and efficiency of medical data collection and sharing. Additional benefits of blockchain within health care are to enable new healthcare-related applications, such as decentralized clinical trials and secure health data storage. Ultimately, the use of blockchain technology in healthcare has the potential to improve patient care while reducing costs and increasing transparency and efficiency across the entire healthcare ecosystem. Challenges that must be resolved to facilitate the widespread adoption of blockchain include addressing concerns with data security and privacy. Other problems plaguing health care that remain challenges for blockchain to overcome include interoperability with existing systems, cost of technology adding to escalating health care costs, scalability, the need for standardization in a highly regulated industry, and the education and training of health care professionals on any emerging technology. Finally, the chapter concludes with prompts for further discussion and application.

Introduction

Medical records are another digital asset that exhibits strong advantages for sharing important medical records among the participants of the healthcare industry to improve patient care. An individual's healthcare records are generally fragmented and stored on the healthcare provider's network, while valuable digital data is available, it is not shared among providers to personal private healthcare records in a manner that facilitates critical medical care. Imagine the possibility of all your medical records being shared on a central technology where all medical professionals treating your condition can access data in real-time, view test results, and add additional commentary to collaborative work towards a compressive solution for your health. Entitling access to sharing medical records on an all-inclusive platform facilitates crucial medical diagnosis (Chen et al., 2019). Collaboration of medical records can save many lives, and the next life that might be saved is yours. Therefore, using a blockchain platform, medical records can be effectively shared with a network of medical professionals at the discretion of the patient.

Big data and IoT data contain consumer data that is at risk due to cybercriminal activity. Securitization of personal private healthcare records to be shared with only authorized providers without the risk of public exposure is a big endeavor and a major undertaking for the healthcare industry. Currently, regulatory agencies and institutions are tasked with regulating big data and security; the task seems gargantuan due to the continued growth in the volume of data making it very difficult to properly secure the data. Garvin (2019) reported that institutions and individuals have been reluctant to store personal electronic medical records due to the cost of setting up an electronic database, fear of the risk of unauthorized access by hackers, and the preset notions of Health Insurance Portable and Accounting ACT (HIPAA) that are restrictive.

What is Blockchain Technology, and How Does it Work?

Blockchain technology is a dispersed ledger system that allows for secure, transparent, and tamper-proof record-keeping (Zheng et al., 2017; Yaga et al., 2019). Every block in the chain holds a cryptographic hash of the previous block, a timestamp, and transaction data. Network nodes verify transactions through consensus algorithms and then record them on the blockchain. When those transactions involve cryptocurrencies like Bitcoin, the ledger is public and decentralized, meaning anyone can view it. Also, a blockchain comprises three essential parts: blocks, nodes, and miners. Blocks are units of data that are chained together—hence the name "blockchain." Nodes are devices that store and transmit blocks throughout the network. Miners are users who validate transactions and add them to blocks. When miners add a block to the blockchain, they receive a small amount of cryptocurrency as a reward.

The "blocks" in the blockchain refer to each transaction, and the blockchain grows as more transactions are added to the ledger. The key advantages of blockchain technology are its security, transparency, and immutability. Because blockchain is a decentralized system that runs on a peer-to-peer network, no central authority can be hacked or corrupted. Additionally, all transactions are transparent and viewable by anyone on the network. Once a transaction is stored on the blockchain, it cannot be changed or deleted. This may sound like a simple concept, but it has significant implications. For example, let's say you wanted to send money to a friend in another country. Traditionally, you must go through a bank or other financial institution to make the transaction. This can be a slow and expensive process because the bank needs to verify the transaction and convert the currency.

With blockchain, you can bypass the bank and send the money directly to your friend. The transaction would be recorded on the blockchain and could not be altered or removed. Plus, since blockchain transactions are peer-to-peer, they are generally much faster and cheaper than traditional bank transfers. Another significant advantage of blockchain is that it is incredibly secure. For a hacker to tamper with the blockchain, they would need to hack into every single computer in the network—no small feat! That's why many believe blockchain is the future of online security. Blockchain technology has a wide range of potential applications, from simplifying supply chain management to streamlining the settlement of financial securities. In the future, blockchain-based systems could potentially revolutionize many industries and aspects of our lives.

Blockchain technology is changing the way we interact with the digital world. By creating a decentralized network of computers to store data, blockchain provides a secure and efficient way to make online transactions. So far, blockchain has most commonly been used for cryptocurrency, but its applications are endless. It's safe to say that we will see more blockchains in the future.

Current Healthcare Framework

1. Initiate Personal Medical Record Retrieval.
2. Perform Medical Record Search.
3. Conform to HIPAA by verifying and upholding HIPAA restrictions.
4. Build Blockchain-Based Database to store medical records (a decentralized, hybrid blockchain that allows the authorized user to access data, provides transparency, and facilitates better patient care through comprehensive diagnosis).
5. Secure data (blockchain application built-in security for each record that is recorded on the chain).

As an individual's lifetime journey continues, the need for medical care changes. When an individual seeks new treatment and seeks care from a specialist that has not previously taken care of them, new data is generated for the patient's healthcare treatment. However, the involvement of several old and new medical professionals and networks creates obstacles in the critical data flow. A common issue that individuals face regarding healthcare is the need to retrieve their medical records when they visit new providers that are involved in their healthcare treatment (O'Keeffe et al., 2021). Another hindrance in data download and sharing occurs because the older population may be intimated when it comes to technology; however, the critical care that the elderly need due to their health issues makes the application of new technology even more serious (Herazo-Padilla et al., 2021). Many medical tracking devices that allow restrictive information regarding the patient such as a person's illness or general diagnosis;

however, may not provide the ability to retrieve complete medical records that are critical for patient care (Wyman, 2020).

What are the Benefits of Using Blockchain Technology in the Healthcare Industry?

You may have heard of blockchain technology concerning cryptocurrency, but did you know that this groundbreaking tech also has the potential to revolutionize healthcare? Here are three ways that blockchain technology could change the healthcare industry for the better.

There are multiple benefits of implementing blockchain-based technology in the healthcare industry. For example, blockchain could help to improve the accuracy and efficiency of medical data collection and sharing. Blockchain-based smart contracts could streamline insurance claims processing and other financial transactions within the healthcare industry. Moreover, blockchain technology could also enable new healthcare-related applications, such as decentralized clinical trials and secure health data storage. Ultimately, the use of blockchain technology in healthcare has the potential to improve patient care while reducing costs and increasing transparency and efficiency across the entire healthcare ecosystem.

Improved Data Security

Data security is among the healthcare industry's most pressing issues today (Haleem et al., 2021). With the 2017 Equifax data breach as a poignant example of 147 million U.S. consumers negatively impacted by their personal information being accessed by unauthorized cyber-attackers, it's no wonder that many people are hesitant to entrust their personal information to healthcare organizations. With so much sensitive information being exchanged electronically, there is a heightened risk of cyberattacks and data breaches. Blockchain technology could help to mitigate these risks by storing data in a decentralized database that is incredibly difficult to hack (Farouk et al., 2020). One of the promising applications of blockchain technology in health care is its ability to reduce fraud and error. In 2021, healthcare expenditures in the United States rose by 2.7 percent, totaling $4.3 trillion or $12,914 per capita. Health spending comprised 18.3 percent of the country's Gross Domestic Product (CMS, 2022, December 15). This represents a lot of money, including billions of dollars in health insurance claims. Some of these claims are fraudulent. Although they only make up a small fraction of all the claims, those fraudulent claims have a very high price tag. They cost us financially and also instill doubt in the trustworthiness of our healthcare system. The National Health Care Anti-Fraud Association (NHCAA) estimates that the financial losses due to healthcare fraud are in the tens of billions of dollars each year. A conservative estimate is 3% of total healthcare expenditures, while some government and law enforcement agencies place the loss as high as 10% of our annual health outlay. This could mean more than $300 billion lost each year because of fraud (Annual Training Conference, 2022).

Enhanced Patient Care

Another potential benefit of blockchain technology is giving patients more control over their health data (Durneva et al., 2020). Patients' medical records are often scattered across different facilities and providers, making it difficult to get a comprehensive view of their health history. With blockchain, patients could access all their health information in one place, leading to better communication between patients and their care providers and ultimately improved patient care. In addition, blockchain technology could also be used to improve patient care. For example, imagine you're admitted to the hospital for surgery. Your medical team will need to order medication and supplies for your procedure—but what if those supplies are out of stock? With blockchain technology, your medical team can track inventory in real-time and order the necessary supplies well in advance—ensuring that you receive the best care.

Improve Interoperability

One of the healthcare industry's significant challenges is the lack of interoperability between systems. Missing interoperability can make it difficult for different providers to access and share patient data—which is crucial for providing coordinated care. Blockchain technology has the potential to change all of that (Dagher et al., 2018).

Reduced Costs

Blockchain technology has the potential to help reduce costs throughout the healthcare system by streamlining processes and eliminating the need for paper records. Blockchain could also streamline claims processing (Angraal et al., 2017). Patients can be reimbursed by their insurance companies for a few weeks or even months. With blockchain, reimbursements could happen in real-time, making life much easier for patients and providers. For example, claims processing is often time-consuming and expensive, but blockchain could make it much simpler and more efficient. This would free up resources that could be redirected to other areas, such as patient care or research and development. For example, consider the claims process. When a patient receives treatment from a healthcare provider, that provider will send a claim to the patient's insurance company for reimbursement. The insurance company will review the claim and either approve or deny it. If the claim is approved, the provider will be reimbursed for the cost of services rendered. However, if the claim is denied, the provider may need to resubmit the claim with additional knowledge—which can result in delays in reimbursement. With blockchain technology, providers could submit their claims directly to the insurance company using a secure, digital ledger. This would eliminate the need for paper claims and speed up the reimbursement process for both providers and patients.

Medical Records

One of the most common applications for blockchain in healthcare is storing medical records (Dubovitskaya et al., 2017). Blockchain can provide a secure, decentralized way to store patient health information (PHI) that is accessible to authorized parties but cannot be modified or deleted. This could potentially eliminate errors caused by human error, such as transcription errors, and make it easier to share medical records between different facilities.

Prescription Drugs

Tracing prescription drugs is another potential use for blockchain in healthcare (Zakari et al., 2022). Blockchain-based drug tracking would allow authorities to trace the origins of counterfeit medications and track legal medicines as they move through the supply chain. This would help ensure that patients receive safe, effective medications and help crack down on illegal drug activity. In 2016, it was estimated that counterfeit drugs accounted for $75 billion in global losses. Blockchain could help by providing a complete record of a drug's journey from manufacturer to patient. This would help ensure that patients are getting safe and effective medications.

Clinical Trials

Blockchain technology can also streamline clinical trials by providing a secure way to store and share patient data (Wong et al., 2019). This would allow researchers to identify eligible patients and speed up recruitment more easily. In addition, it would enable trial sponsors to keep track of patient payments and improve compliance with regulations.

What Challenges must be Addressed before the Widespread Adoption of Blockchain Technology in Healthcare?

Before blockchain technology can be widely adopted in healthcare, several challenges need to be addressed. These include ensuring data security and privacy, fraud and misuse of medical data, privacy concerns, interoperability, and lack of standardization, regulation, scalability, and cost.

Ensuring Data Security and Privacy

The key concern around adopting any new technology is data security and privacy (Marr, 2022). With healthcare data being susceptible, any blockchain solution must ensure that patient data is securely stored and cannot be accessed or tampered with without permission. Healthcare data is sensitive and needs to be protected. Even with blockchain, there are data storage and potential security concerns that need to be addressed before its widespread use in healthcare (Morrissey, 2020, August 18). This issue must be resolved for blockchain technology to be adopted on a widespread basis in health care.

Fraud and Misuse of Medical Data

One of the most significant risks associated with blockchain technology is the potential for fraud and abuse of medical data (Morrissey, 2020, August 18). With so much sensitive patient information being stored on the blockchain, there is a real danger that hackers could gain access to this data and use it for nefarious purposes. To combat this risk, it is essential that only trusted, and authorized individuals have access to the blockchain.

Privacy Concerns

Another significant risk associated with blockchain technology is the potential for privacy breaches. If patient information is stored on the blockchain, there is a risk that unauthorized individuals could access this information. Only authorized individuals must have access to the blockchain to protect patient privacy.

Interoperability

Another challenge that needs addressing is the ability of different systems and devices to work together. This is important in healthcare as many different systems and databases are used, often from various vendors. A blockchain solution must integrate seamlessly with existing systems to be adopted on a broader scale. For blockchain technology to be truly effective in healthcare, it must be able to interoperate with existing systems. Otherwise, it will create silos of data that other systems cannot access. This issue must be resolved before blockchain technology can be widely adopted in healthcare.

Lack of Standardization

One of the biggest challenges facing the adoption of blockchain technology in healthcare is the need for more standardization. Currently, many different platforms and protocols are vying for attention, each with advantages and disadvantages. For blockchain technology to be widely adopted in healthcare, there needs to be an agreement on which platform or protocol to use. Otherwise, the various platforms and protocols will end up competing instead of working together.

Regulation

Health care is a heavily regulated industry, and any new technology needs to comply with existing regulations. Blockchain solutions will need to be designed in such a way that they meet all relevant regulatory requirements. Regulatory uncertainty is the biggest challenge facing the adoption of blockchain technology in health care (Stanley, 2018). Because blockchain technology is still relatively new, there must be clear guidelines for regulating it. This creates a risk for both investors and companies looking to adopt blockchain technology. Until there is more clarity on how regulators intend to deal with blockchain-based systems, it is unlikely that we will see widespread adoption of this technology in health care soon.

Scalability

Scalability is one of the challenges facing any new technology—the ability to meet increasing demand as more users adopt it. Blockchain solutions will need to scale up seamlessly to be adopted on a broader scale. Blockchain networks can only process a few transactions per second. This is insufficient to meet healthcare needs, generating a large amount of data. For blockchain technology to be adopted on a widespread basis in healthcare, the scalability issue must be addressed.

Cost

Another challenge that needs addressing is cost. Healthcare is already a costly industry, and any new technology needs to be cost-effective to be adopted on a broader scale.

Education and Training

The challenge that needs addressing is education and training. As with any new technology, blockchain solutions must be accompanied by educational materials and training for healthcare professionals to ensure their successful adoption.

What is the Future of Blockchain Technology in the Healthcare Industry?

The future of blockchain technology in the healthcare industry is shrouded in potential but fraught with uncertainty. The hope is that blockchain will help to address many of the systemic problems plaguing healthcare, such as interoperability, data security, and fraud prevention. In the meantime, healthcare organizations should closely monitor blockchain developments and experiment with the technology where it makes sense. With suitable applications, blockchain could help to transform healthcare for the better.

Relevant Blockchain Case Studies

1. **MedRec:** MedRec is a blockchain-based medical record system designed to give patients control over their medical data (Azaria et al., 2016). It allows patients to create, manage, and share their medical records with healthcare providers. The system ensures the accuracy and privacy of medical data by using cryptographic techniques to secure the data.
2. **Tierion:** Tierion is a blockchain-based platform designed to improve the security and transparency of medical data (Tierion: Blockchain Proof Engine, 2023). The platform allows healthcare organizations to create verifiable

records of their data, ensuring that the data is accurate and tamper-proof.

3. **Patientory:** Patientory is a blockchain-based platform designed to improve patient engagement and data sharing in healthcare. The platform allows patients to create, manage, and share their medical data with healthcare providers securely and in real-time, improving the accuracy and efficiency of patient care.

4. **Guardtime:** Guardtime is a blockchain-based platform designed to improve the security and integrity of medical data. The platform uses blockchain technology to create an immutable record of medical data, ensuring that the data is accurate and tamper-proof.

5. **MedicalChain:** MedicalChain is a blockchain-based platform designed to give patients control over their medical data. The platform allows patients to create, manage, and share their medical records with healthcare providers securely and in real-time, improving the accuracy and efficiency of patient care.

MedRec Case Study

MedRec is a Blockchain-based electronic medical record system developed by researchers at the Massachusetts Institute of Technology (MIT) (Azaria et al., 2016). The system uses Ethereum Blockchain technology to store medical records, and is designed to provide patients with complete control over their medical records, including who can access them. The development of MedRec is a significant step towards enhancing the security and privacy of medical records while enabling transparent and secure data sharing between doctors and patients.

One of the primary benefits of MedRec is its emphasis on privacy and security. In traditional medical record systems, patient data is stored in centralized databases that are vulnerable to hacking and other forms of cyberattacks. MedRec, on the other hand, uses Blockchain technology designed to be secure and tamper-proof. Medical records are stored on a decentralized ledger, and patients can grant access to their records to healthcare providers only if they approve. This provides patients with greater control over their medical records, and it enhances the privacy and security of patient data.

Another significant benefit of MedRec is that it can be used for clinical trials. Clinical trials require access to large amounts of medical data, and MedRec provides a secure and transparent mechanism for data sharing between doctors and patients. With MedRec, patients can choose to share their medical records with researchers and other healthcare providers securely, ensuring that sensitive medical information is not compromised. This can enhance the accuracy and validity of clinical trials and contribute to the development of new and more effective medical treatments.

MedRec is also designed to be interoperable, meaning it can work with other electronic medical record systems. This is critical for healthcare providers, as many organizations use different electronic medical record systems, making it difficult to share patient data across different systems. MedRec's interoperability feature allows healthcare providers to access patient records in real time, regardless of which electronic medical record system is used. This streamlines the sharing of medical information and enhances the accuracy and efficiency of medical treatment.

However, there are also some challenges associated with MedRec's adoption. One significant challenge is regulatory compliance. Healthcare organizations must comply with various regulations, including HIPAA and GDPR, which dictate how patient data must be handled and protected. MedRec is a relatively new technology, and regulations surrounding its use in healthcare are still in their infancy. As a result, healthcare organizations must carefully evaluate the regulatory implications of implementing MedRec before it can be fully adopted.

Another challenge is the complexity of implementing MedRec. Blockchain is a complicated technology that requires specialized knowledge and expertise to implement effectively. Healthcare organizations may need to invest significant resources in training and hiring staff with the necessary skills to design, develop, and maintain MedRec.

In conclusion, MedRec is a Blockchain-based electronic medical record system that has the potential to revolutionize the way medical data is stored and shared. MedRec provides patients with greater control over their medical records, enhances the privacy and security of patient data, and streamlines the sharing of medical information between healthcare providers. While there are some challenges associated with its adoption, MedRec's potential benefits make it an exciting development in the healthcare industry, and it can improve the accuracy and efficiency of medical treatment while protecting patient privacy and security.

Tierion Case Study

In today's world, healthcare data is one of the most valuable assets, and the security and transparency of this data are crucial for ensuring the safety and privacy of patients. Tierion is a blockchain-based platform designed to improve the security and transparency of medical data. The platform provides a tamper-proof, decentralized system for storing and sharing medical data, which can help to control data breaches and safeguard patient privacy.

A key feature of Tierion is its capability to create a tamper-proof record of medical data. The platform utilizes blockchain technology to create a secure, transparent, and immutable record of all medical data, including patient records, test results, and treatment plans. This ensures that medical data is always up-to-date and accurate and that any changes to the data are recorded in a transparent and auditable way.

In addition, Tierion provides a range of tools for securely sharing medical data. The platform uses a decentralized network of nodes to ensure that data is always available, even if some nodes go offline. This helps to prevent data loss and ensures that medical data is always accessible when it is needed. Another important feature of Tierion is its ability to verify the authenticity of medical data. The platform uses cryptographic algorithms to ensure that data has not been changed or tampered with, providing a high degree of security and trust. This can help to control data breaches and protect patient privacy, which is crucial in today's world where cyber-attacks are becoming increasingly common.

In general, Tierion is a promising solution to the challenges faced by the healthcare industry in securing and sharing medical data. By leveraging the unique features of blockchain technology, the platform can provide a secure and transparent system for storing and sharing medical data, which can help to improve patient outcomes and protect patient privacy.

Patientory Case Study

Patientory is a blockchain-based platform that is designed to improve patient engagement and data sharing in healthcare (Your Health at Your Fingertips, 2023). The platform provides patients with a secure and transparent way to manage their health data, giving them a more active role in their healthcare. Patientory has the potential to revolutionize the way patients interact with the healthcare system, ultimately leading to better patient outcomes and reduced healthcare costs.

One of the key features of Patientory is its ability to store and share patient health data securely. The platform uses blockchain technology to build a tamper-proof and transparent patient data record, which healthcare providers can easily access and share. This ensures that patient data is always up-to-date and accurate, which can lead to better treatment outcomes.

In addition, Patientory provides a range of tools for patient engagement, including personalized health goals, health trackers, and patient communities. These tools enable patients to take a more active role in their healthcare, empowering them to make informed decisions about their health and well-being. By improving patient engagement, Patientory can help to reduce the risk of complications and improve patient outcomes. Another critical feature of Patientory is its focus on interoperability. The platform is designed to work with various healthcare systems, enabling patients to access and share their health data across different providers easily. Interoperability can help to reduce the risk of errors and improve patient outcomes by ensuring that healthcare providers have access to all relevant information about a patient's health.

Largely, Patientory is a promising solution to the challenges faced by the healthcare industry in engaging patients and sharing health data. By leveraging the unique features of blockchain technology, the platform can provide a secure, transparent, and patient-centered solution to healthcare data management and patient engagement. As the healthcare industry continues to evolve, Patientory is likely to become an increasingly important player in the quest to provide better care at a lower cost while improving patient outcomes.

Gaurdtime Case Study

Guardtime is a blockchain-based platform designed to improve medical data security and integrity. The platform provides a tamper-proof, decentralized system for storing and sharing medical data, which can help to control data breaches and protect patient privacy. Guardtime has the potential to revolutionize the way medical data is managed and shared, ultimately leading to better patient outcomes and reduced healthcare costs.

One of the critical features of Guardtime is its capacity to create a tamper-proof record of medical data. The platform utilizes blockchain technology to create a secure and transparent record of all medical data, including patient records, test results, and treatment plans. This ensures that medical data is always up-to-date and accurate and that any changes to the data are recorded in a transparent and auditable way. In addition, Guardtime provides a range of tools for securely sharing medical data. The platform uses a decentralized network of nodes to ensure that data is always available, even if some nodes go offline. This helps to prevent data loss and ensures that medical data is always accessible when it is needed.

Additionally, an important feature of Guardtime is its ability to verify the authenticity of medical data. The platform uses cryptographic algorithms to ensure that data has not been changed or tampered with, providing a high degree of security and trust. This can help to control data breaches and protect patient privacy, which is crucial in today's world where cyber-attacks are becoming increasingly common.

On the whole, Guardtime is a promising solution to the challenges faced by the healthcare industry in securing and sharing medical data. By leveraging the unique features of blockchain technology, the platform can provide a secure and transparent system for storing and sharing medical data, which can help to improve patient outcomes and protect patient privacy. As the healthcare industry continues to evolve, Guardtime is likely to become an increasingly important player in the quest to provide better care at a lower cost while ensuring the privacy and security of patients' data.

MedicalChain Case Study

MedicalChain is a blockchain-based platform that is designed to give patients control over their medical data. The platform provides a secure and transparent system for storing and sharing medical data, which can help to improve patient outcomes and give patients more control over their healthcare.

A key feature of MedicalChain is its ability to give patients control over their medical data. The platform allows patients to securely store their medical data on the blockchain, where it can be accessed by healthcare providers when needed. This can help to ensure that patients have access to their medical data whenever they need it, and that they can control who has access to their data. In addition, MedicalChain provides a range of tools for securely sharing medical data. The platform uses a decentralized network of nodes to ensure that data is always available, even if some nodes go offline. This helps to prevent data loss and ensures that medical data is always accessible when it is needed.

A distinct feature of MedicalChain is its ability to verify the authenticity of medical data. The platform uses cryptographic algorithms to ensure that data has not been changed or tampered with, providing a high degree of security and trust. This can help to control data breaches and protect patient privacy, which is crucial in today's world where cyber-attacks are becoming increasingly common.

MedicalChain is an all-embracing, promising solution to the challenges faced by patients in managing their medical data. By leveraging the unique features of blockchain technology, the platform can provide a secure and transparent system for storing and sharing medical data, which can help to improve patient outcomes and give patients more control over their healthcare. As the healthcare industry continues to evolve, MedicalChain is likely to become an increasingly important player in the quest to provide better care at a lower cost while ensuring the privacy and security of patients' data.

Healthcare Privacy Issues

Several authors have concluded that privacy issues create fear among individuals and make them uncomfortable regarding the idea of storing and retrieving personal health care data on the internet which could face the risk of unauthorized access and cybercriminal attacks, making the data vulnerable (Kumar and Shantala, 2020; Posey et al., 2017). Healthcare generates data that is complex to aid in the prevention, diagnosis, and treatment of diseases related to various healthcare issues. Currently, big data is the compilation of many medical data points, and HIPPA regulations create issues around masking the privacy of the user. It is reported that building trust between patients and adopting new technology is critical to advancing medical data sharing (Antes et al., 2021; Dwivedi et al., 2019).

The demand for critical care among the patient population is rising and so is the concern for data privacy and security and the need for sharing critical medical data points. Technology evolution has created sophisticated wearable devices that are revolutionizing and contributing to the advancement of healthcare by assisting healthcare professionals to monitor critical data about the patient (Zarowitz, 2022). The healthcare industry is adapting to technological advancement and embracing emerging technology to push innovation in the industry. Adept patient data is contributing to a better-informed patient population. Advancements in healthcare innovations have reached exponential growth and resulted in healthcare treatment and improved quality of life for patients (Vervoort et al., 2022). While the push towards emerging technology is much recognized in the healthcare industry, current technology is not successful in achieving data-sharing capability efficiently. Thus, the industry is grappling with access to stored data regarding advanced medical diagnosis that allows medical practitioners and patients to retrieve critical information safely, securely, and smartly.

The COVID-19 pandemic created a much-needed push toward the adoption of emerging technologies to facilitate patient healthcare in ways that were not accessible. For instance, the use of Telemedicine, Tele-screening, and Tele-ophthalmology to harness technology to create a touch-free environment while still providing critical care to the population. Many authors support the premise of digital transformation applications and technology innovations in the healthcare industry and believe that the innovations are expected to grow exponentially in the next few years (Barakat et al., 2017; Li et al., 2021) especially as the pandemic continues and threats of a new wave of diseases in the future.

HIPAA regulates and defines the standards for transmitting healthcare records electronically. HIPAA legislation was passed in 1996 and ensures oral, written, and well-documented methods that enhance and ensure healthcare record privacy, security, and confidentiality. At the heart of the policymakers' agenda was to streamline healthcare record storage and protect consumer information. Since its inception, HIPAA has exposed vital and deeper healthcare issues for policymakers. The core objective of HIPAA is to protect patients' confidentiality of healthcare records, and healthcare privacy and safeguard patients against inadvertent privacy information disclosure (Alfandre et al., 2020). HIPAA has captured the interest of healthcare industry providers, networks, and various other entities to pursue electronic record safety (Strauss, 2018).

Application of Emerging Technology

Complex medical diagnosis such as cancer requires a different approach to data sharing. In a study conducted by de Vocht and Roosli (2021), it was revealed that caring for cancer patients requires psychosocial and physical factors while in-patient care in the hospital. Patients with terminal, incurable diseases often must engage in a complex healthcare treatment that has to be managed through personalized psychosocial factors of various support elements to provide a sense of safety, calm, and hope. All medical and supporting factors that are shared among the patients' medical providers are an avenue for better healthcare for the patient. However, personal health records have experienced an incessant need to protect private healthcare data and active sharing between providers. Many healthcare databases currently are operating on non-rigid schema conformity which means that they do not support relational table modifications and sharing of data efficiently among users (Hoque and Hoque, 2018). Personal healthcare records are composed of many different types of datasets that are structured, unstructured, or semi-structured such as social security, birth records, health insurance information, various health diagnostic records from providers, immunization status, specialized test results, etc. (Razmak and Belanger, 2018). A blockchain technology-based data-sharing system among medical professionals' networks and authorized users facilitates data visibility for patients, family members, medical consultants, researchers, and regulatory agencies such as the Center for Disease Control and Prevention (CDC). Blockchain provides a robust solution for electronic data sharing securely to facilitate the next generation of advantages for patient care.

Healthcare Privacy Perceptions

The use of Information and Communication Technologies (ICT) is prevalent in the healthcare industry. According to a study by Cherrez-Ojeda et al. (2020), many healthcare professionals depend upon ICT to search for reputable publications to enhance patient treatment. Data access on mobile platforms is also very attractive to users as it provides faster data transfer, with low-cost and easy access to medical information (Tang et al., 2019). Sjostrom et al. (2022) concluded that the adoption of communication technologies has improved patients' perception and acceptance and eased technological barriers. The healthcare industry has a huge opportunity to utilize blockchain technology to access other technologies to merge onto one platform and provide access to all stakeholders in a controlled, safe environment to protect the patient's identity while facilitating collaboration and efficacy. Complexity with electronic prescriptions and reorders is currently subpar. Electronic prescriptions (e-RX) allow healthcare providers to communicate the prescription needs of a patient electronically. There are many parties involved in the e-RX process that need to be secured and validated. In addition, providing secure access to patients to share their prescription-related information elevates the level of patient care. Blockchain technology can validate, secure, and share data among all participants that have access to the platform creating substantial proficiencies for both individuals and healthcare providers. Ansari et al. (2022) postulate that the COVID-19 pandemic has emphasized a pronounced socioeconomic gap in healthcare delivery and accessibility. This gap in healthcare delivery and accessibility can be addressed through digital technology such as blockchain that provides a large storage network for data securely, inexpensively, and competently. The exponential growth of telemedicine has created an access explosion where it is estimated that 93% of individuals are poised to access telehealth (Da Lomba, 2022). The explosion of telemedicine requires healthcare organizations to adopt end-to-end encryption, which creates an exciting next chapter for the healthcare industry to upgrade to blockchain technology for data storage and distribution.

Structure

Blockchain technology is appealing to organizations as it promises real-world solutions (Bamakan et al., 2021). A blockchain is a technical approach to data storage and data sharing that creates a reliable database. Identification of types of transactions that are industry specific can be stored on the blockchain platform and any related information is recorded in different parts of the blockchain. As users desire to identify and manage datasets or build relationships the blockchain database can successfully facilitate these transactions. Gopalakrishnan et al. (2021) demonstrated that information regarding specific wastes is shared with organizations that allow the organizations to track a particular type of waste product that may be of interest to them. Kassou et al. (2021) assert that the financial industry can greatly benefit from blockchain technology.

The rise in interest in data security to protect patient confidentiality, launched the adoption of various technologies to build the regulations into healthcare management and healthcare supply chain management processes (Kim and Hyun Jun, 2019; Wyman, 2020). The process of electronic medical records (EMR) encompasses many aspects. EMRs include healthcare information such as patients' healthcare reports, appointment details, and information about health conditions, diagnoses, tests, and reports from various providers (Butler, 2018). Digitization of massive amounts of paperwork to electronic healthcare tracking systems provided utility to all users by making it easy to share data. While this is wide-ranging information that involves several institutions, the compilation of this critical mass of data assists medical professionals in better decision-making capabilities for the patient. The challenge that is faced by the industry is how to effectively store the EMR and create measures of data security to comply with the law. Safeguarding large amounts of data presents challenges with current technological systems.

Adapting to new technology has decreased healthcare administration costs and improved patients' quality of care tremendously. Chen et al. (2019), posit that the preservation of data integrity and protection against data vulnerability and corruption is particularly important, especially, since the data exchanges from one system to another. Furthermore, preserving sensitive healthcare information regarding certain diagnosis rates higher, such as, properly protecting mental health diagnosis, or human immunodeficiency virus (HIV) treatment status for patients. Clearly, data security, integration, and strong sharing features are needed in the healthcare industry to provide data privacy protection. Blockchain technology provides a structure that has the capability of storing massive amounts of data securely, providing traceability, and sharing the data effectively.

Besides, a strong need for data privacy, rampant cybercriminal activity creates an existential threat to the healthcare industry. Big data evolution despondently produces real threats of hacking into personal healthcare information that is on the rise and safeguarding the information is the primary concern (Garvin, 2019). Cybercriminal activity has been on the rise in the healthcare industry. In 2020, large healthcare organizations including Blue Cross, Allscripts, and Centers for Medicare and Medical Services fell victim to ransomware, phishing attacks, and several other data breaches. As cyberattacks continue to surge, so is the expected proactive spending to defend against these attacks which are estimated to exceed $65 billion in the next 4–5 years (Garvin, 2019). Integration of healthcare systems on a blockchain-based system creates an elevated level of security, minimizes the risk of hacking, and improves data-sharing capabilities among approved users. Healthcare organizations can gain huge benefits from blockchain integration. A major objective for corporate security is to enable maximum security protection for all assets of the organization. The purpose of corporate security is to respond to threats and proactively detect and defend against attacks. The new evolution of healthcare record management on a digital platform allows bad actors to breach data security. Patients are one of the important new users of the new digital platform. Cybercriminals can easily attack unsuspecting individuals trying to navigate a new technology through the deployment of social engineering schemes.

Social engineering is the use of trickery to influence individuals to divulge personal confidential information that can be used for fraudulent purposes. Algarni et al., 2020 maintain that social engineering is a means for cyber attackers to gain access to organizations' resources without authorization by deploying tricks that lure the victims using phishing emails, online chats, SMS, etc. Individuals are more susceptible to falling victim to social engineering tricks especially since the digitization of medical records is an extremely new phenomenon for the masses. Cybercriminals take advantage of the lack of knowledge of individuals and exploit them to access sensitive, confidential information to inflict serious damage with malicious intent (Lee et al., 2017b). Unsuspecting users are targeted by hackers to expose vital data that is then used to gain access to the entire network and spread malicious malware restricting the system, giving hackers full control of the network (Zainudin, 2021). Likewise, social engineering is indisputably an effortless means to gain illegal access to sensitive organizational data (Gonzalez-Zarzosa and Diaz, 2021). A large survey of data conducted in the United Arab Emirates healthcare system revealed that the privacy of patients' electronically held records creates data security issues that are believed to jeopardize patient care data integrity and are responsible for eroding trust between patient-doctor relationships (Bani et al., 2020).

Blockchain technology creates an enormous opportunity for the metamorphosis of personal health records. The current state of the healthcare industry is struggling with data security and the vast potential of data sharing in a secure manner. While the industry is vigilant about the changes and has embraced technology for solutions, there is a huge opportunity to use blockchain applications to standardize healthcare industry initiatives. Personal health records are presently stored in stand-alone systems that operate in a controlled centralized environment. Conventional technology relies upon old file system technology with limited portability, poor access control structure, and non-encrypted technology (Hassan et al., 2021). The power of IoT, medical smart devices, information systems explosion, and various cloud-based services have led to a digital explosion and transformation of the healthcare industry, but this digital explosion has also contributed to data breaches in the sector (Seh et al., 2020). In addition to platform data breaches, advanced healthcare devices that track patient data have empowered healthcare users while also providing easy access to online services based on the premise of digital healthcare and wallet-based systems propagating external attacks and unauthorized access.

The healthcare industry is divided into large organizations. Ateetanan and Shirahada (2018) uphold that larger organizations with bigger resources should engage in the competitive market to create a strategic advantage by adopting innovative technologies as a full-service solution. Furthermore, because of their study, Southworth Davis et al. (2018) reported that organizations and consumers have been waiting for innovative technologies to clear the adoption and security gap and provide simpler technological capabilities and interfaces.

While initially both the healthcare industry and the patients were skeptical of advanced technological applications and showed resistance, a study conducted by several authors has substantiated that the industry and the patients are in favor of implementing new technologies (Antes et al., 2021; Man Lai et al., 2019).

Framework

Healthcare can create a future that improves the quality of life of a citizen and the collective productivity of a nation. It is the most talked about topic across every election cycle on how to cater to the needs of an evolving demographic that is constantly moving, diverse, and digitally savvy. The disparity in doctor to patient ratio between developed and emerging nations are eye-opening ranging from 2.4 per thousand in developed countries to 0.2 per thousand in emerging countries, technology and specifically a decentralized model that enables trust in the provider and the citizen, and more importantly, creates and derive the value of the data created that benefits to the community at large. With these various drivers, the framework for evolution must bridge the service and technology as one integrated whole. Hence, we propose a framework defined below to enable the future of healthcare (see Figures 4, 5, and 6).

VIRTUALISATION

Deviceless, virtualized primary healthcare services as a virtual experience with embedded AI

DATAFICATION

Federated data aggregation creating patient 360. Asset value unlocked for data benefit. Patient data ownership & privacy ensured.

TOKENIZATION

Digitalization of all assets for global value exchange and equitable distribution of value

Figure 4: Framework Healthcare Industry
(Source: Sashi Edupuganti, ODE Holdings)

Virtualization: While the pandemic had a lot of challenges, some benefits came out of it. One among those is the acceptance of telemedicine or digital engagement as people were not encouraged to go to the hospitals and now it has become a norm. This creates new opportunities for enabling "doctors across borders" by virtualizing digital healthcare across the city, state, and national boundaries without physical limitation. The ability to reduce the healthcare gap by decreasing the doctor-per-patient ratio can be and only be enabled by a trusted technology that blockchain enables.

Datafication: Data is the new oil that fuels human healthcare by proactively monitoring, collecting, predicting, and easing healthcare across continents. The key to this is creating a Medical360 that is a holistic dataset of the user across the

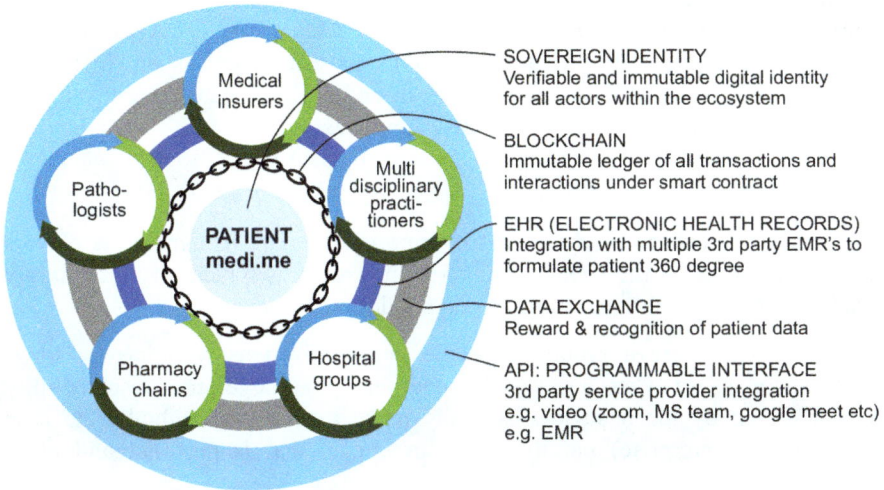

Figure 5: Blockchain framework: Healthcare Industry
(Source: Sashi Edupuganti, ODE Holdings)

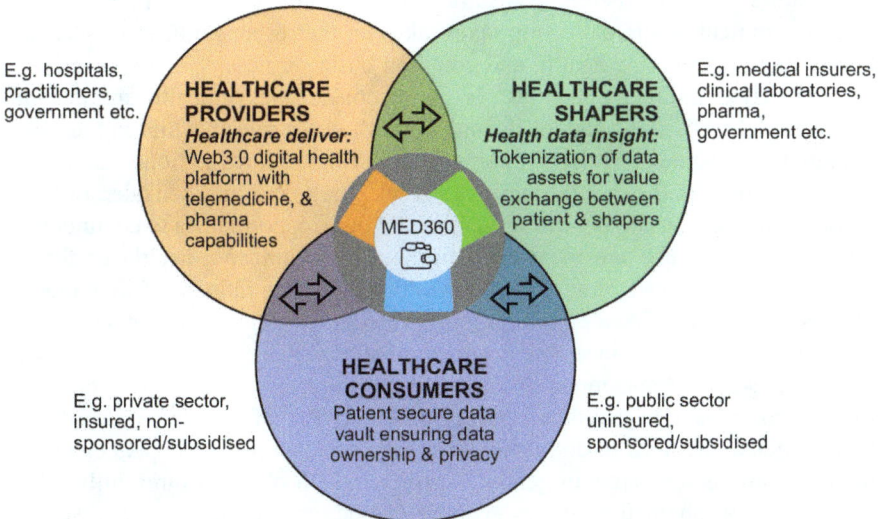

Figure 6: Blockchain application: Healthcare Industry
(Source: Sashi Edupuganti, ODE Holdings)

Note: This framework was created by ODE Holdings, Inc as a part of their vision for a Healthcare subsidiary focused on creating a new future of healthcare.

healthcare ecosystem, for the first time we can create a decentralized ecosystem that creates and supports individual health management.

Tokenization: Healthcare data is the most important in the digital world, in the dark web, healthcare data fetches $250 per record as compared to its closest rival financial card information which fetches $6. The value of such data while protecting the privacy of individuals can be unleashed to fuel the digital transformation of a nation. This can and will be enabled by tokenizing Medical360 for individual and collective monetization.

Ecosystem

For any framework to work well, it is important that we recognize the players and the personas in the supply chain. Healthcare, as is typical of other supply chains, has at least three prominent players described below.

Shapers: These are a group of people or organizations who have a long-term vision as to what benefits the community at large, this responsibility is not driven by the need to essentially make a profit (it might be the case in a private enterprise) but to create an impact that is personal and close to them.

To a very large extent, this role is played by government organizations which hasn't always been the case as private enterprises step in when the government lacks the innovation/research muscle. The example of SpaceX having to stand up a shuttle program when NASA stalled is an example of where it has worked. The government, while collecting taxes takes on the responsibilities of security, human rights, economic health, and common basic services.

Directly or indirectly, they take on the responsibilities of healthcare, in some cases as in the UK or Canada it is done through direct means while in the US it is largely a private enterprise supported by the rules and regulations enabled by the government at the federal and state level. They take into consideration the current challenges, changing demographics and other opportunities to continuously propose and introduce new means and mechanisms to influence healthcare for the nation at large. The shaping aspect of healthcare can be influenced by strategic players who see a business opportunity to accelerate the innovation that has largely not been able to be provided by the government or existing players. In healthcare today the likes of Amazon exploring an opportunity or insurance players trying to push for proactive healthcare are prime examples of shapers who can and will introduce the need to do this strategically. An example of shapers changing the landscape is the "Human genome" project which decoded three billion bits of information about the human genome to understand and prevent diseases. These projects are typically done with a focus on technology, value network, and innovative opportunities. Shapers through innovation have an opportunity to save at least 100 billion dollars in inefficiencies that can be directed to more productive and proactive management of healthcare.

Providers: These are existing individuals, groups, or organizations providing the services as a part of their business model. In the case of healthcare, these would be doctors, lab technicians, pathologists, hospitals, nutritionists, etc. They are a part of creating, optimizing, and evolving the services to provide value to their customers. These entities are thinking midterm and have annual plans on business goals that drive the focus and energy. Especially in healthcare, these organizations are slowly moving given the number of providers is fewer in number and the cost of entry and exit is very high. Innovation for them exists mostly at the iterative level and rarely you do see leapfrogging due to the pork that exists. The adoption of blockchain and Web 3.0 is going to be based purely on operational savings for the most part.

Consumers: These are individuals who need services related to their health, reacting to an existing issue or being proactive in maintaining a healthy lifestyle.

Application

Organizations have struggled with system security concerns. Blockchain decentralized and symmetrically distributed ledgers provide several security functionalities in a decentralized setting by utilizing cryptographic systems (Odeh et al., 2022). Largely scattered and fragmented data now have an opportunity to be centralized in a blockchain database that fundamentally provides robust system security thereby creating value for organizations.

Blockchain applications provide wide-ranging applications in many industries such as finance, healthcare, manufacturing, education, etc. Blockchain technology offers trustworthiness, transparency, and credibility to data storage. The core properties of blockchain technology are decentralization, immutability, and transparency. Blockchain is a point-to-point distributed network where no single third party is involved during the communication and transaction creation process.

Distributed Ledger Technology—blockchain is a peer-to-peer network that operates on the premise of distributed ledger technology where users on the blockchain digitally verify the transactions without the intervention of a trusted third-party authority. In distributed ledger framework main benefits are immutability, security, distributed, programmable, unanimous, and anonymous.

Blockchain has a wide-ranging array of applications as it provides solutions for addressing security concerns surrounding data as the main component. On a blockchain, the ledger that is the fabric on which the blockchain is constructed is a compilation of transactions. Decentralized ledgers provide access to multiple parties in an organization. This allows users to access the distributed ledgers and obtain identical versions of the distributed data on the blockchain. Alterations made to the ledger such as editing, deleting, and adding are validated and reproduced to all participants who have ledger access to maintain symmetry and accuracy of the data in a synchronized manner.

Data Security

A blockchain functions on the premise of asymmetric cryptography. Asymmetric cryptography uses a double-layer approach of a public and private key to access the blockchain database. The public key is used for encrypted access and the private key is used for decryption. Asymmetric keys in a blockchain authenticate identities for both transactions and personnel.

Anonymity is a distinctive feature of the blockchain. Participants on the blockchain do not disclose their identities, instead, data is stored through asymmetric cryptography that provides identification such as an address for digital signatures where the users sign the transaction digitally with their private key while creating the transaction. This feature allows users of the blockchain to authenticate transactions without disclosing their identity.

Transactions are the data flow that transmits value and code between different accounts using a smart contract policy. Anti-spoofing is achieved when each transaction on the blockchain is signed digitally using an asymmetric digital signature crypto algorithm.

Some challenges in using blockchain are current market conditions and obstacles to user-centric approaches in industries. This is mainly caused by the old infrastructure and leaders are unable to fully understand the concept of blockchain technology to move their organization's data management into the next generation.

Several authors have pointed out issues with legacy systems and issues surrounding the adoption of blockchain technology. Ratta et al. (2021), Rahmani et al. (2022), and Bigini et al. (2021) all concur that blockchain technology adoption is greatly beneficial to several industries; however, the potential obstacles to blockchain adoption are dependent upon adopting a user-centric approach supported through the elimination of legacy systems to consolidate data in a blockchain-based database. In addition, some of the legacy system storage, especially ones that utilize cloud computing have struggled with centralization, enormous overhead, trust evidence, reduced adaptability, and inaccuracy.

The integration of blockchain technology provides benefits to all users especially consumers who can access and share data creating time and money efficiencies. Andoni et al. (2019) conducted a thorough literature review that revealed that many practical applications of blockchain technology are fueled by industries needing to provide an atmosphere where two different parties can trust each other. In addition, Andoni et al. also documented the limitations of blockchain adoption and concluded that no mention of the wider application within industries currently exists.

Challenges

Some challenges with blockchain are lack of adequate standardization, shortage of trust between parties, and integration of blockchain technology from legacy systems. Odeh et al. (2022) assert that while several industry standards that are followed industry wide, a core way of conducting business is developed in legacy frameworks that require login and passwords for each user, cloud storage, and different databases to store information that creates fragmentation. Organizations need to lean on their leadership to understand the benefits of blockchain technology to develop streamlined systems and processes to move away from legacy systems and grow their businesses into a platform that allows the push to the next generation.

Conclusion

Personal private healthcare records are often fragmented and stored on different healthcare providers' networks, which makes it difficult for authorized providers to access and share critical medical information. This can result in fragmented care and can negatively impact patient outcomes. Additionally, big data and IoT data contain valuable consumer information that is at risk of cybercriminal activity, and securing personal healthcare records to be shared only with authorized providers without the risk of public exposure is indeed a major undertaking for the healthcare industry. Regulatory agencies and institutions are responsible for regulating big data and security, but as mentioned, the volume of data continues to grow, making it challenging to secure the data properly. However, there are ongoing efforts to improve data security and privacy in healthcare, such as the adoption of standardized electronic health records and the implementation of data encryption and secure data-sharing platforms. These efforts aim to ensure that personal healthcare information is kept safe and accessible only to authorized individuals.

Blockchain is a decentralized database that uses cryptography to secure and verify transactions. Transactions are stored in blocks that are linked together in a chronological and immutable chain. Once a block is added to the blockchain, it cannot be altered or deleted without consensus from the network. This makes blockchain a reliable and tamper-proof database. Industry-specific transactions can be stored on the blockchain platform, and any related information is recorded in different parts of the blockchain, making it easy to track and verify the entire history of the transaction. This is particularly useful in industries such as finance, where there is a need for secure and transparent transactions. In addition, blockchain can be used to facilitate data management and build relationships between users. For example, a supply chain management system can use blockchain to track the entire journey of a product, from the raw materials used to manufacture it to the final delivery to the customer. This can help increase transparency and accountability in the supply chain. Thus, blockchain technology has the potential to revolutionize industries by providing a secure, transparent, and efficient way to store and share data.

The preservation of data integrity and protection against data vulnerability and corruption is particularly important, especially since the data exchanges from one system to another. Furthermore, preserving sensitive healthcare information regarding certain diagnosis rates higher, such as, properly protecting mental health diagnosis, or human immunodeficiency virus (HIV) treatment status for patients. Clearly, data security, integration, and strong sharing features are needed in the healthcare industry to provide data privacy protection. Blockchain has a wide-ranging array of applications as it provides solutions for addressing security concerns surrounding data as the main component. On a blockchain, the ledger that is the fabric on which the blockchain is constructed is a compilation of transactions. Decentralized ledgers provide access to multiple parties in an organization. This allows users to access the distributed ledgers and obtain identical versions of the distributed data on the blockchain. Alterations made to the ledger such as editing, deleting, and adding are validated and reproduced to all participants who have ledger access to maintain symmetry and accuracy of the data in a synchronized manner.

Chapter Summary

Medical records contain crucial information about a patient's health history, ongoing treatments, medications, allergies, and other important details that are essential for providing effective and coordinated healthcare. However, due to the fragmented nature of healthcare systems, these records are often stored in different locations and formats, making it difficult for healthcare providers to access and share this information. Sharing medical records electronically can improve patient care by providing healthcare providers with a more complete picture of a patient's health history, reducing the risk of medical errors, and improving coordination between healthcare providers. Electronic health records (EHRs) and health information exchanges (HIEs) are two technologies that facilitate the sharing of medical records among healthcare providers. However, there are also concerns about the privacy and security of medical records, as they contain sensitive and personal information. Therefore, it is important to ensure that appropriate safeguards and regulations are in place to protect the privacy and security of patient's medical records while also enabling the sharing of this information for improved patient care.

Blockchain technology has the potential to disrupt various industries and has already been used for various purposes beyond just digital currency transactions. The security of the blockchain is indeed one of its key features, as its decentralized structure makes it difficult for hackers to manipulate the data. The immutability of blockchain also makes it appealing for use cases where data integrity is critical, such as in supply chain management, voting systems, and land registries. With its potential to increase transparency and efficiency while reducing costs, blockchain technology is likely to continue to play an important role in shaping the future of many industries.

The recent pandemic has indeed accelerated the adoption of virtualization technologies in healthcare. Telemedicine, remote consultations, and digital healthcare have become increasingly popular in recent times, and this trend is likely to continue in the future. Virtualization enables doctors to reach patients across different locations, irrespective of geographical boundaries, making it possible for patients to access quality healthcare services even from remote areas. Blockchain technology, with its inherent security features, can be leveraged to create a trusted and secure platform for virtualizing digital healthcare. By enabling the secure sharing of patient data across different healthcare providers and stakeholders, blockchain technology can significantly reduce the healthcare gap and improve patient outcomes.

Datafication, or the process of converting healthcare data into actionable insights, is another important trend in healthcare. With the increasing availability of healthcare data, it is now possible to create a comprehensive Medical360 dataset that captures the patient's complete health journey across different healthcare providers and ecosystems. This dataset can be used to proactively monitor and predict patient health outcomes, identify potential health risks, and enable personalized healthcare interventions. In conclusion, virtualization and datafication are two important trends in healthcare that are transforming the way we access, deliver, and manage healthcare services. By leveraging technologies like blockchain, we can create a secure and decentralized ecosystem that empowers patients, healthcare providers, and other stakeholders to collaborate and improve patient outcomes.

Take Away Lessons from the Chapter

1. An individual's healthcare records are generally fragmented and stored on the healthcare provider's network, while valuable digital data is available, it is not shared among providers to personal private healthcare records in a manner that facilitates critical medical care.
2. Big data and IoT data contain consumer data that is at risk due to cybercriminal activity. Securitization of personal private healthcare records to be shared with only authorized providers without the risk of public exposure is a big endeavor and a major undertaking for the healthcare industry.
3. The "blocks" in the blockchain refer to each transaction, and the blockchain grows as more transactions are added to the ledger. The key advantages of blockchain technology are its security, transparency, and immutability.
4. A blockchain is a technical approach to data storage and data sharing that creates a reliable database. Identification of industry specific types of transactions can be stored on the blockchain platform and any related information is recorded in different parts of the blockchain.
5. Organizations have struggled with system security concerns. Blockchain decentralized and symmetrically distributed ledgers provide several security functionalities in a decentralized setting by utilizing cryptographic systems.
6. Distributed Ledger Technology—blockchain is a peer-to-peer network that operates on the premise of distributed ledger technology where users on

the blockchain digitally verify the transactions without the intervention of a trusted third-party authority. In distributed ledger framework, the main benefits are immutability, security, distributed, programmable, unanimous, and anonymous.
7. A blockchain functions on the premise of asymmetric cryptography. Asymmetric cryptography uses a double-layer approach of a public and private key to access the blockchain database.

Chapter Exercises

1. What are at least three ways of using blockchain-based technology that would benefit the healthcare industry?
2. What are the key challenges to using blockchain technology in the healthcare industry?
3. What suggestions would you make if consulting for leaders in the healthcare industry in support of considering blockchain?

Blockchain Applications and Uses: Law Enforcement

Blockchain technology was introduced with the advent of digital currency. Regulators had limited knowledge of the product and even more limited regulations to manage the criminal activities that followed the illicit use of the new system. The precipitous use of digital currency has increased criminal activities such as money laundering, insider trading, fraud, and other financial crimes. Global governance and regulators are struggling with digital asset regulation. The concept of digital assets was the first of its kind. In addition, limited knowledge and non-existent regulatory framework unintended consequences led to cybercriminal activity that is currently prevalent. Blockchain technology can limit cybercriminal activity and facilitate checks and balances for regulators to monitor. The advantages of blockchain technology include expanding access to safe and affordable financial services and reducing the cost of domestic and cross-border funds transfers and payments transforming the modernization of public payment systems. The technological architecture of digital assets has substantial implications for data privacy, national security, and operational security and resilience of financial systems. Two main factors of law enforcement that can be facilitated through the application of blockchain technology: accountability and justice. In the criminal justice system, the chain of custody, decentralized identification management, data sharing and management, and authentication of body camera footage can all be achieved through effectively collaborating through a common set of principles and guidelines to assist law enforcement.

Introduction

Stagnant laws are fueling illegal activities surrounding digital currencies. Deterring illegal use of digital assets by building a regulatory framework to support digital currencies, regulators can mitigate risks, exploits, crashes, and thefts and provide current regulators like the SEC and CFTC oversight to protect the consumer. Digital currency is widely used to pay for goods and services, and investments. While the intent of creating digital currency at heart was to facilitate a decentralized currency that would be attractive to the user, digital currency has delivered a myriad of financial crimes. Since the advent and precipitous use of digital currency has increased, so has the unlawful activity of money laundering, insider trading, fraud, and other financial crimes instigated by underdeveloped laws to monitor and regulate digital currency. In 2008, a person or a group of people published a white paper, Bitcoin: A Peer-to-Peer Electronic Cash System under the pseudonym Santoshi Nakamoto. Bitcoin emerged in 2009 on the Genesis Block and started trading online in 2010 (Duggan, 2022). As digital assets these were the first of their kind to be introduced; however, as it is with new inventions, there is initially limited market knowledge about the innovation and non-existent regulations to monitor the unintended consequences of it. The introduction of digital currency was no different. Regulators had limited knowledge of the product and even more limited regulations to manage the criminal activities that followed the illicit use of the new system.

The SEC and CFTC can play a significant role in regulating the digital currency market. They can provide oversight and ensure that the market is fair and transparent. They can also enforce regulations to prevent illegal activities and punish those who engage in them. In addition, they can work with other regulatory bodies to develop a coordinated approach to regulating the digital currency market. The lack of regulations in the digital currency market has created a breeding ground for illegal activities (Tsai, 2021). To mitigate these risks, regulators must develop a regulatory framework that supports digital currencies while deterring criminal activity. The SEC and CFTC can play a significant role in this process by providing oversight and ensuring that the market remains fair and transparent. Ultimately, a well-regulated digital currency market can provide significant benefits to consumers while deterring criminal activity.

Although global governance and regulators are struggling with digital asset regulation, here in the US, the government is focusing on new legislation that will impact digital assets. Hussein and Sweet (2022) suggest that lawmakers are grappling with pointing primarily to regulatory authority over digital assets, developing regulations, laws, and enforcement. Green (2022) upholds that current laws regarding digital currencies are in proposal format and have both Bank Secrecy Act (BSA) and US Patriot Act at their center. In addition, The Financial Crimes Enforcement Network (FinCEN) is another agency that operates under the Treasury Department to monitor illicit criminal activity by enforcing BSA regulations. Under the current Biden Administration, The White House released an Executive Order (EO) in March 2022 to address the framework and interagency coordination concerning digital assets. The extensive EO release by President Biden pointed to policy construction, objectives, coordination, and policy and actions to protect consumers, investors, and businesses in the United States. The Executive Order released by President Biden in March 2022 highlights the government's commitment to addressing digital asset regulation comprehensively. The order's focus on policy construction, coordination, and protection of consumers, investors, and businesses indicates that the US government is taking a proactive approach to digital asset regulation. The order recognizes that digital assets' growth has profound implications for data privacy and security, and the government must take steps to mitigate potential risks.

The EO cited that advancements in digital and distributed ledger technologies have led to the exponential growth of digital assets that has profound repercussions for the protection of consumers, investors, and businesses encompassing data privacy and security. Digital asset regulation is becoming an increasingly important issue for governments worldwide, and the US is no exception. The US government is actively working to develop new legislation and regulations to govern digital assets and protect consumers, investors, and businesses. The Bank Secrecy Act, US Patriot Act, and the Financial Crimes Enforcement Network are all critical components of the US government's efforts to combat illicit activity in the digital asset space. As the digital asset space continues to evolve, the US government will likely continue to refine its regulatory framework to ensure that it keeps pace with these developments. Digital asset market participants must stay up to date with regulatory changes and comply with any new requirements to avoid running afoul of the law.

The advantages of blockchain technology include expanding access to safe and affordable financial services and reducing the cost of domestic and cross-border funds transfers and payments, transforming the modernization of public payment systems. The main objectives of the EO are consumer, investor, and business protection against cybercriminal activities and to protect human rights abuses through the enforcement of laws.

Illicit finance and national security risks generated by the misuse of digital assets can be mitigated through several measures, including:

1. **Implementing robust Know Your Customer (KYC) and Anti-Money Laundering (AML) measures:** Digital asset service providers should implement robust KYC and AML measures to ensure that their services are not being used for illegal activities. This can include verifying the identity of customers, monitoring transactions, and reporting suspicious activities to the relevant authorities.

2. **Regulating digital asset service providers:** Governments can regulate digital asset service providers to ensure that they comply with AML and counter-terrorist financing (CTF) regulations. This can include licensing requirements, mandatory reporting of suspicious activities, and sanctions for non-compliance.

3. **Enhancing international cooperation:** Illicit finance and national security risks generated by the misuse of digital assets are global issues. Governments can enhance international cooperation to share information and coordinate efforts to combat these risks.

4. **Increasing public awareness:** Public awareness campaigns can help raise awareness about the risks associated with digital assets and the importance of reporting suspicious activities to the relevant authorities.

5. **Developing new technologies:** New technologies, such as blockchain analytics, can help detect and prevent illicit activities involving digital assets. Governments and the private sector can work together to develop and implement these technologies.

By implementing these measures, governments and the private sector can help mitigate the risks associated with the misuse of digital assets and protect the financial sector from illicit activities such as money laundering, cybercrime, ransomware, narcotics, terrorism, and human trafficking.

The EO confirms that promoting access to safe and affordable financial services to all Americans is of strong interest and achievable through responsible innovation that has traditionally underserved Americans due to inequitable cost-efficiency access. The technological architecture of digital assets has substantial implications for data privacy, national security, operational security and resilience of financial systems. Meanwhile, achieving efficiencies in the traditional banking system to facilitate cheaper, faster, and safer transaction execution by promoting greater and more cost-efficient access to financial products and services is desired. The United States Central Bank Digital Currencies (CBDC) exists to enact policies and actions related to digital currencies.

Money laundering has been a prevalent issue for digital currencies. In their work, Sadon (2021) discussed the process of digital currency money laundering. Sadon (2021) described five forms of digital currency money laundering. Nested Services take advantage of more than one exchange. For instance, using decentralized finance (DeFi) the exchange provides buying and selling side transactions. Including an over the counter (OTC) broker where the transaction can pass through the OTC broker's exchange and then to a DeFi exchange is an example of a nested service. Gambling platforms are used to place bets and then cast out winnings legitimizing the money transaction anonymously. Mixers blend digital assets by sending assets back to new addresses or digital wallets. Non-compliant exchanges do not regulate or follow compliance regulations to identify and verify the transfer of digital assets. Organizations operating in high-risk jurisdictions appeal to criminals because lack of jurisdiction facilitates cybercriminal activity without any recourse. Technology platforms are deficient in anti-money laundering efforts because KYC protocols are violated by criminals who manipulate the system using pseudonyms or anonymity.

Structure

Sai et al. (2023) uphold that equitable finance is not commonly available due to current infrastructure availability in the traditional financial market. Sai et al. (2023) believe that a viable solution is the application of blockchain technology to decentralize KYC and reduce multiple KYC verification to create access to microcredit to benefit disadvantaged segments of the population. Financial assistance is needed to promote financial development among underdeveloped societies. However, the banking structure at its core is designed to lend against fixed income and collateral. Two principles negate lending for individuals who lack traditional bank lending requirements. Underprivileged consumers desperately need tools that can prevent financial barriers and open financial lending for them.

The adoption of digital technologies has created innovative solutions to penetrate traditional financial institution requirements. For instance, one innovative solution is the maintenance of databases where lender and borrower information has been documented effectively. At the core, many borrowers are unsuccessful in borrowing funds due to the mishandling of documents and inaccessibility of documentation. In addition to mishandling the database for informational purposes, many employees in the sector are not properly trained, which creates more manual work and generates more human error. Untrained workers and accessibility create a time-consuming and untraceable environment where consumers encounter a lack of confidence and trust. Blockchain technology has successfully penetrated many industries such as finance and supply chain. Traditional databases that record transactions have created traceability and trust vulnerabilities for the consumer and industry alike. The foundation of blockchain technology is corner-stoned in mechanisms that provide a trust-free system to ensure data privacy and security in a decentralized manner where access is available to all users on the blockchain. The blockchain technology framework ensures trust, transparency, privacy, security, and reduces fraudulent activities, and thereby increasing productivity.

The genesis of blockchain technology develops began with a protocol developed by David Chum in 1982. Two mathematicians, Stuart Haber, and W. Scott Stornetta implemented a system of tamper-resistant timestamps that created the basis for the chain. Haber incorporated a merkle tree or hash tree in the design proposed by Stornetta and Bayer in 1992. Finally, in 2008 Satoshi Nakamoto released a white paper focusing on a decentralized blockchain, which was implemented later that year through the generation of Bitcoin.

A distributed ledger that stores data digitally is the primary function of a blockchain. The data on a blockchain is collected in units of blocks, and each block contains data sets. After a block is finalized, it is contained and linked to the block prior to it, thereby creating a blockchain. Blocks have predefined storage capabilities. Thus, after the first block is formulated, every single new bit of information is merged into a new block, which is then added to the chain.

Traditional databases store data in tables, while blockchain stores data in blocks that are linked together. When one block is filled up to capacity, it is time-stamped and added to the chain. This provides an exact order of events thereby adding a layer of transparency. Each new block is interconnected to the previous block through the previous block's hash and its timestamp. The blockchain database spreads the data among several nodes in different locations, this helps redundancy and preserve the integrity of the data. A copy of the chain exists in each node of the blockchain network that is updated as soon as new blocks are added. Blockchain allows anyone to see live transactions made on the blockchain and this function provides trust at the heart of the technology. The mechanism continues to protect the data and facilitate trust because if there are unsavory characters that access the blockchain database and make fraudulent changes, the copy gets singled out because it will no longer match all other copies while cross-referencing. For instance, if a malicious user who has somehow accessed a node on the blockchain wants to make changes to certain transactions on the blockchain, that user will have to dominate and make changes to more than half of all the copies of the blockchain. For the malicious user to accomplish dominating changes to more than half of the copies of the blockchain there is a massive amount of work, resources, and computing power needed; therefore, it is extremely difficult to compromise a blockchain database. This creates immutability and integrity for the database.

Casino et al. (2019) assert that the financial sector would gain the most from incorporating blockchain technology because the technology provides fast output and benefits in the traditional banking sector through an increase in productivity. The use of smart contracts is an important part of blockchain architecture. In the early 1990s, Nick Szabo introduced smart contracts, a concept of a set of promises in a digital format that includes protocols among parties that perform on these promises. A smart contract is designed to carry out business logic in a decentralized setting through a reaction of incidence. When the blockchain fabric is constructed, developers design smart contracts by incorporating the terms and conditions required by the application. For instance, the exchange of money, delivery of services, and the release of digitally restricted content. Smart contracts can also be utilized for enforcing privacy protection, such as releasing select data that is privacy-protected to comply with a specific request. Essentially, the primary function of smart contracts is to carry out business logic programmatically and execute a particular task, process, or transaction in response to a specific set of criteria.

Know Your Customer (KYC) is a procedure to obtain confidential client information pertaining to the identity of clients in the financial sector. KYC is critical to the financial sector because it allows investment banking decisions in the best interest of the client. Since the information collected is confidential, the financial sector is more predisposed to illegal activity. Information collected for KYC purposes can be stored on a blockchain as it is a permanently unalterable design called self-sovereign identity. A self-sovereign identity system is governed by the premise that every individual must have control over the administration of their identity. Therefore, users can control their data through a self-sovereign identity database on a blockchain. In addition to maintaining identity, users also have control over what data they'd like to share or keep private. The self-sovereign blockchain database allows users to authenticate, delete, and maintain their data and eliminates the need for bureaucratic processes to verify identity.

The financial sector needs a database technology design that increases efficiency, trust, and transparency by considerably decreasing human interactions and removing third parties from the transaction. The financial industry's core business model is lending, which is dependent upon establishing secure peer-to-peer connections between the financial institution and the consumer to store transactional data in a secured database (Agi and Jha, 2022). The safety and security of lending, investments management, and borrowing are all transactions based. In a blockchain, all these transactions can be captured to avoid manipulation or deletion of the records.

The need for *accountability* and *justice* is paramount for law enforcement, especially the criminal justice system. The possibility of using blockchain with its core capabilities of immutability, and consensus are a great enabler for driving these needs for accountability and justice.

Chain of Custody: According to the National Registry of Exonerations (2023) there have been 3381 exonerations from 1989 with more than 28,770 years lost. Out of these numbers, a total of 772 cases were due to tampered evidence, this happens to be about 23% of all exonerations. Imagine people losing decades of their lives because of misplaced evidence or corruption. Among many organizations that fight for these lives and opportunities lost is the *Innocence Project, Equal justice initiative, Raices, Urban Justice Center, and Legal Aid*. The intention of law enforcement and criminal justice is to minimize this with a level of transparency and change public opinion. This is important not just for the many lives impacted by the wrongful judgments but also for the need for law enforcement to reinvent itself as a transparent and reliable organization to avoid the likes of defunding law enforcement, which while sounding great will have a direct and deep impact on the community at large. We need a healthy law enforcement organization for us to live the lives we so desire.

To help this transformation new frameworks have been developed, the most important of which being the Alister, Inc framework for Blockchain for Evidence (BOE). This framework created a new supply chain that established the *chain of custody*. This framework, while nascent, will focus on creating a network of entities and individuals in the law enforcement ecosystem who have access to the evidence while capturing the data and metadata about the touch point. Each touch point is consented to and recorded by multiple parties including the public making the touch point an immutable record for perpetuity. When this evidence is called for to be presented, the validation of these immutable records beyond the walls of the evidence hall will provide transparency and integrity to the evidence minimizing or eliminating the evidence-related impact in cases, especially in the criminal justice system. The usage of this tamper-proof architecture might not fit the traditional use case of blockchain as it relates to tokenization but that part of the value of using a blockchain would favor.

1. **Decentralized identity management:** There are roughly 18,000 law enforcement agencies in the USA alone, with a ratio of about 2.4 sworn officers per 1000 citizens (Community Certified Policing Services, 2015). That is a lot of agencies and officers, and the number goes even more when you include civilian officers. That ratio becomes 3.4 per 1000 citizens. While the numbers by themselves are not scary given that it is mostly in line with

the number of doctors we have per citizen ratio. Among other things, what is scary is the level of coordination and cooperation required by the agencies, which are mostly decentralized with some progress for centralization or effectively collaborating through a common set of principles and guidelines.

2. **Data sharing and management:** In law enforcement, there is a need for a tremendous amount of data and intelligence sharing, some of it for a short period of time validated by multiple parties. Therefore, the current legacy systems do not facilitate optimal data sharing and management because there is no formal communication through database sharing between the different agencies.

3. **Authentication of body camera footage:** The use of blockchain technology for body camera footage enables accountability, security, and transparency building trust between the public and law enforcement organizations. Due to its inherent characteristic of immutability, the use of blockchain creates a tamper-proof system. Furthermore, the distributed nature of storage ensures that the video footage is available at multiple locations for ease of use and accountability, eliminating the dependency on a node to manage or "release" the footage as it is deemed appropriate. With the evolution of blockchain from smart contracts to secret contracts and its public nature, the level of security that is available including encryption can and will possibly be used for increased privacy protection of individuals. Trust becomes a by-product of enabling camera footage through blockchain.

4. **Smart contracts:** The availability and use of smart contracts will possibly eliminate the need for manual dependency and processing of transactions between various departments in law enforcement through automation. This highly accountable and accountable architecture will minimize bureaucracy and the risk of errors that exist with existing centralized and manual processes and systems.

Framework

The left quadrant of Fig. 7 outlines the current framework for county, state, and federal agencies and regulators. Presently each government agency operates in a silo with a very limited amount of information shared among the entities. Data fragmentation, manual paperwork, lack of technological innovation, and lack of collaboration are all factors that have severely impacted government agencies and created inefficiencies that increase the cost. For instance, if a consumer were to request their local municipality for a transaction, the trial for that transaction is limited to the municipality and not shared among all other agencies. The same applies if the same consumer were to transact at the state level, the information is stagnant in the state database. Law enforcement needs a technology that can leapfrog through decades of neglected updates to their system. As depicted in the framework, all agencies can effectively communicate and share data through the use of blockchain, as modeled on the right quadrant of Fig. 7.

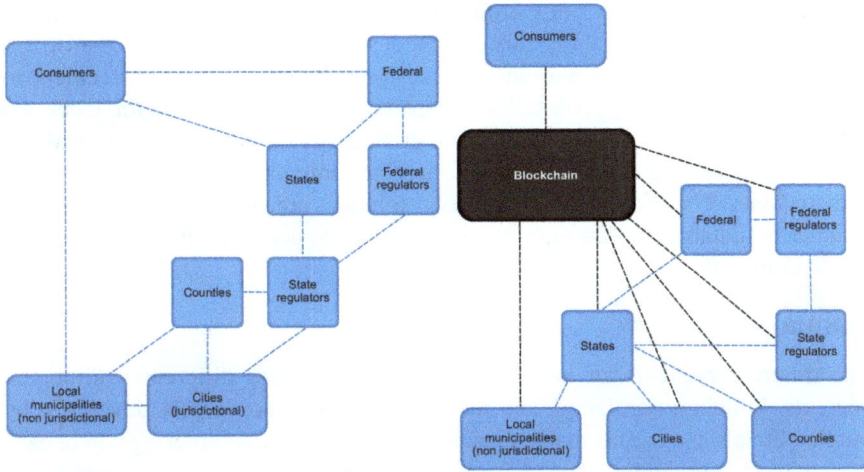

Figure 7: Framework Law Enforcement Industry
(Source: Author's own work)

Note: The framework for the governance of data across several government agencies is shown in Figure 5 and the clear impact of blockchain technology to create a collaborative venue to share data among these agencies.

Data Fragmentation

As clearly exhibited by the framework shown in Fig. 7, each government entity operates as its own siloed agency. There is little to no communication between each agency, and this is by design to contain and secure data by the agency itself. The creation of law enforcement, regulatory, and compliance agencies by nature has been to collect data for their respective agencies. However, in the digital age, there is more data collected regarding individuals, their transactions, and their purchasing habits that have become a critical part of forensic data for law enforcement. Traditionally law enforcement agencies have worked within familiar silos which are easier to manage and may be less resource intensive while working across boundaries and collaboration intensifies opportunity costs, productivity, and cost efficiencies. In law enforcement, the push towards collaboration has led to elevated expectations of core competencies and the desire for better public safety. For instance, emphasizing the importance of increasing the visibility and accessibility of law enforcement is to reassure the public of effective policing. The objective of law enforcement is to ensure the following:

* Reduce fear of crime
* Increase public confidence in law enforcement agencies
* Increase community efficacy
* Improve intelligence gathering
* Reduce anti-social behavior
* Reduce crime

The tasks outlined for law enforcement agencies are critical to public and national safety; however, the inclusion of thousands of agencies just in the United States alone creates a conundrum and an uphill battle for law enforcement to lead with intelligence. Nevertheless, there is an immediate need for local and national law enforcement agencies to integrate closely to facilitate analytical and planning processes and establish and form complimentary partnership structures and processes. Therefore, blockchain-based solutions are optimal for a traditionally fragmented database of law enforcement agencies to connect with each other and create efficacies.

Manual Paperwork

The different branches of law agencies play out across different institutional contexts. Policing, regulators, and the judicial system are largely developed by public officials and are normally seen as a doming of private ordering where each agency works autonomously and is not designed to communicate with each other. Currently, if there is a police report that needs to be filed, it must be manually written by the police officer on a piece of paper and assigned a report number. The manual report is then filed at the local police station and the assigned report number is then used to track any changes to the report in the future. Imagine a world where a police report is generated electronically by a police officer that is directly loaded into the blockchain system and becomes available near real-time to all agencies. If the victim of the crime chooses to get a restraining order from the court, the court automatically pulls up the information and authenticates the police report using the unique identifier and the court issues the restraining order upon request. Once the restraining order is approved, all agencies have access to the information on the said document. There is huge potential for blockchain applications that removes manual paperwork, human error, and redundancies and creates harmony among all users in the most effective manner.

Lack of Innovation

Policymakers are undergoing tumultuous debates about synchronizing big data for the use of law enforcement. The orthodox view that policymakers employ currently excludes all shareholders' equity regarding data sharing. The debate is arguing that access should be broadened to consider the interests of all stakeholders, such as citizens, the environment, and communities (Kovvali, 2022). Facilitating the visibility of data among law enforcement agencies is needed to address important social problems and to create them if left unchecked. There is an ostensible inability of the political system to create better solutions and the flexibility of the law to address problems. A shared toolkit approach is needed, especially in light of the recent COVID-19 crisis that not only began to erode trust and demand virtual presence, but staunchly showcased the many deficiencies with the old legacy approach used by law enforcement agencies. The rise in digital information and environmental changes demands that there is innovation in the framework for law enforcement agencies. As shown in Fig. 7, the data is managed in silos by each agency. The incorporation of blockchain technology will not only decentralize the system and remove redundancies, but it will also create effective and efficient collaboration among the agencies that will lead to the possibility of better public welfare.

Lack of Collaboration

Blockchain technology has the potential to enhance collaboration among law enforcement agencies by providing a secure and transparent platform for sharing information and data. One of the key features of blockchain is its ability to create a tamper-proof and decentralized ledger that can be accessed and updated by authorized parties in real time. By leveraging blockchain technology, law enforcement agencies can create a secure and transparent system for sharing information such as criminal records, arrest warrants, and other sensitive data. The decentralized nature of the blockchain ensures that information is not controlled by any single entity and cannot be easily tampered with or altered without leaving a trace.

Moreover, blockchain technology can also enable the creation of smart contracts, which can automate the execution of certain processes, such as the sharing of information or the execution of arrest warrants. Collaboration can help reduce the administrative burden on law enforcement agencies and streamline their operations, leading to greater efficiency and faster response times. Definitively, the use of blockchain technology can also help build trust among law enforcement agencies and other stakeholders, as it provides a transparent and verifiable record of all actions taken on the platform. This can help improve collaboration and cooperation among different agencies and promote greater accountability and transparency in law enforcement operations.

Furthermore, law enforcement agencies can work collaboratively to solve cybercriminal activities. In fact, collaboration is often necessary to effectively investigate and prosecute cybercrimes, which can involve multiple jurisdictions, countries, and even continents. Collaboration can take many forms, including information sharing, joint investigations, and joint task forces. It may involve cooperation between local, state, federal, and even international law enforcement agencies, as well as with private sector partners, such as cybersecurity firms and financial institutions. Effective collaboration requires clear communication, trust, and a shared understanding of each other's roles and responsibilities. It also requires a willingness to share resources and expertise and to work towards a common goal of bringing cybercriminals to justice.

While cybercrime can be complex and challenging to investigate, collaboration among law enforcement agencies can improve the chances of identifying and prosecuting those responsible. Overall, while the implementation of blockchain technology in law enforcement may face some challenges, such as interoperability between different systems and the need for standardized protocols, its potential benefits in enhancing collaboration and improving efficiency make it a promising avenue for further exploration.

Application

Preferential partnering persists in inter-organizational boundaries within government agencies. Complex intractable public safety problems have increased multi-organizational and multisector collaborations. Building inter-organizational networking is an active approach to attaining resources, sharing knowledge, and improving outcomes for the safety of the public (Bevc et al., 2015). To understand preferential tendencies or silos, the structural relations among the organization's members must be evaluated. Silos impede upon efficiencies brought forth by collaboration such as cohesiveness.

There is a push towards a futuristic approach to policing that is led by strategy, collaboration, and intelligence to control crime. Governments are moving towards greater efficiency and effectiveness in public services and actively deploy intelligence-led policing to achieve public safety (Maguire and John, 2006). In 2004, National Intelligence Model (NIM) started operating at the heart of policymaking in England and Wales. The purpose of NIM was to ensure mainstream policing that utilizes strategic and futuristic frameworks to approach crime control (Hale et al., 2005; Maguire, 2000; Maguire, 2003; Maguire, 2004). There are many factors that were considered while creating the pilot for NIM, such as continued frustration with the central government regarding public safety.

An active need to develop better investigation and evidence gathering to avoid encountering problems created by using more traditional reactive methods is needed. Traditional reactive methods of investigation and evidence gathering often involve waiting for an incident to occur before acting, which can result in missed opportunities to prevent or mitigate harm, as well as difficulties in gathering evidence that is crucial to understanding what happened and who is responsible (NTT DATA, 2022). To avoid encountering these problems, it is important to develop better investigation and evidence gathering methods that are proactive and preventative. This can include:

- Establishing early warning systems and protocols that enable organizations to identify potential risks before they escalate into full-blown incidents.
- Conducting regular risk assessments to identify potential vulnerabilities and develop strategies to mitigate them.
- Using data analytics and other tools to proactively monitor and analyze large volumes of data to identify patterns and trends that may indicate potential problems.
- Investing in training and development for investigators and evidence gatherers to ensure that they have the skills and knowledge needed to use new technologies and techniques effectively.

By adopting these and other proactive measures, organizations can improve their ability to identify and address potential risks before they become problems, while also enhancing their capacity to gather and analyze evidence in a timely and effective manner. This can help to reduce the impact of incidents, protect the reputation of the organization, and enhance overall safety and security.

Increased awareness of proactively approaching crime-solving by creating a collaborative network means promoting the idea of actively seeking out and sharing information among different stakeholders to prevent and solve crimes. This can involve many individuals and organizations, including law enforcement agencies, community groups, local businesses, and residents. The idea behind creating a collaborative network is that by pooling resources and information, a more comprehensive and effective approach to crime-solving can be achieved. For example, local businesses and residents may be more likely to report suspicious activity if they feel they are part of a larger effort to combat crime, and law enforcement agencies can use this information to better understand and respond to criminal activity in a given area. Some ways to promote this type of collaboration might include hosting community meetings or workshops to discuss crime trends and prevention strategies, creating online platforms or social media groups to share information and resources, or partnering with local businesses and organizations to implement community-based crime prevention initiatives. Ultimately, the goal of promoting increased awareness and collaboration in crime-solving is to create a safer, more connected community where everyone is invested in preventing and solving crime.

Furthermore, technological advancements utilize sophisticated alternatives to surveil, store, and retrieve large amounts of digital data. The distinction of the NIM model was to elaborate on the outcomes of community safety, reduced crime, arrested/disrupted criminals, manage hot spots and control potentially dangerous offenders through intelligence, and proactive operations. The breakdown of the model assumes three tiers. Level 1 is local area policing or basic command units. Level 2 is regional governance, and Level 3 is national and international threats. Therefore, the tasks and coordination at each level of the NIM model are essential and informed decision-making at each level is not possible if the information is not shared cohesively among all three levels. At the basic level of each group, tasking and coordinating groups comprise of managers that allocate appropriate resources based on intelligence and shared data. For example, a Sheriff Lieutenant will typically deploy resources such as detectives and police officers and include the help of other agencies that have deep knowledge about the events and is not necessarily limited to the Sheriff's department.

There are four intelligence classifications: strategic assessments, tactical assessments, target profiles, and problem profiles. The strategic level of intelligence contains information that is updated about challenging problems and analysis. For instance, if the US government establishes an audit for US Threat Assessment, this will be considered a Level 3 strategic assessment. Tactical assessments manage and control strategies to propose review actions. Tactical assessments also draw attention to emerging problems that require resource allocation. Target profiles and problem profiles are operational and depend upon intelligence on specific criminals or crimes. Law enforcement intelligence is based upon nine analytical techniques resulting from the analysis, crime pattern analysis, market profile, demographics and social trend analysis, criminal business profiles, network analysis, risk analysis, target profile analysis, and operational intelligence assessment. Four modes of intelligence have been identified: criminal intelligence, crime intelligence, community intelligence, and contextual intelligence (Innes et al., 2005). Recently, there has been a revolution to debunk the police in the US due to grave injustices carried out by law enforcement agencies. The public demands reassurance of corrective action and initiatives for more transparent and ethical policing platforms. Law enforcement is a highly visible industry that is being stretched to move into the next century of evolution. Paper trails, manual work, and hidden agendas no longer have a place in law enforcement and due to its massive traditionally manual legacy systems, law enforcement urgently needs a next generation solution to catapult its old operations systems. There is a clear and fundamental shift taking place towards a philosophy of policing built around risk assessment and management that needs trust and transparency at its core. Blockchain-based solutions cater to this core competency.

Digital Asset Case

In February 2022, The Department of Justice (DOJ) announced that two people in Manhattan, NY were allegedly charged with the conspiracy to launder stolen cryptocurrency from a Bitfinex hack in 2016. Around August 2016, Bitfinex, a cryptocurrency exchange, was hacked and approximately 72 million dollars in digital currency was stolen. According to Assistant Attorney General Kenneth A. Polite, Jr. of the DOJ criminal division, federal law enforcement followed the money through blockchain to apprehend the criminals (Department of Justice, 2022). While regulators are keenly seeking an understanding of digital currency, unscrupulous activity seems to run rampant. In August of 2022, the DOJ released that they were extraditing Alexander Vinnik to the United States for prosecution. Vinnik, who is accused of Russian Cryptocurrency money laundering, allegedly laundered approximately $4 billion in digital currency (Department of Justice, 2022). While anonymity is fueling bad actors to commit serious financial crimes, blockchain technology is useful while tracking cryptocurrency based on public ledger technology.

Blockchain-based technology has the capability of tracking transactions based on data stored on the chain. Regulators need to consider training for and gaining access to blockchains where digital assets reside and create regulations to provide law enforcement agencies to be active users with the authority to track transactions using the blockchain technology features itself. It is worth noting that digital currency has been around since the discovery and hyper-popularity of Bitcoin in 2008–2009. Cybercriminals have had an undisputed advantage to manipulate new technology for their benefit. However, executive orders from the world's leading economy, the United States, have only been enacted as of 2022. An approximate 14-year gap needs to be bridged between the preexisting frameworks surrounding digital assets and proposals for law enforcement and regulation of the technology.

Conclusion

Technology is the catalyst for change, innovation, and new business growth. The advent of new technologies has become tools for illicit activities to avoid detection. For instance, digital forensics is needed in law enforcement to modernize investigations and sustain the chain of custody for digital assets. The chain of custody refers to the chronological documentation or paper trail that records the seizure, custody, control, transfer, analysis, and disposition of physical or digital evidence. It is a crucial part of the legal process that ensures the reliability and authenticity of the evidence presented in court. The chain of custody is vital because it establishes a clear and unbroken link between the evidence and the person who collected it. It helps to prevent tampering, loss, or contamination of the evidence, which could compromise its integrity and admissibility in court. A well-documented chain of custody ensures that the evidence is handled properly and can be traced back to its source at any point in time.

The chain of custody is especially important in criminal cases where physical evidence such as DNA, fingerprints, or blood samples are used to link suspects to a crime scene. It is also important in civil cases where the evidence could impact the outcome of the case. Therefore, maintaining the integrity and authenticity of the evidence is critical in ensuring that justice is served, and the chain of custody is a vital component of the legal process that helps to achieve this goal. Blockchain has been an anticipated and reliable technology to provide immutability and traceability for digital content.

The objective of law enforcement is to reduce fear of crime, increase public confidence in law enforcement agencies, increase community efficacy, improve intelligence gathering, reduce anti-social behavior, and reduce crime. The framework included in this chapter focuses on the reduction of the current fragmentation of law enforcement databases that highlights manual paperwork, lack of innovation, and lack of collaboration at the core of the problem.

Indeed, the application of blockchain technology can bring significant benefits to the legal system, particularly in streamlining communication and information sharing among different law enforcement agencies. By leveraging the blockchain, police reports can be securely and immediately shared among agencies and the judiciary, reducing errors and delays in processing legal matters. Additionally, using blockchain technology for legal documentation can increase transparency and accountability in the legal system, as it enables a tamper-proof and verifiable record of legal activities. This can be particularly useful in cases where the authenticity of legal documents is contested or where there is a need to track the history of changes to legal records.

The lack of regulations and oversight in the digital currency market has created a breeding ground for illicit activities. Criminals have taken advantage of the anonymity and lack of transparency that digital currencies offer to carry out illegal activities such as money laundering, terrorist financing, and drug trafficking. These activities harm the integrity of the financial system and can harm innocent consumers who unknowingly participate in these schemes. To address these issues, regulators must develop a regulatory framework that supports digital currencies while also deterring criminal activity. This framework should be designed to protect consumers and provide oversight to ensure that the market remains fair and transparent. It should also include measures to prevent money laundering and other illicit activities. We have provided a comprehensive framework in this chapter that exclusively highlights the integration of blockchain technology in the law enforcement sector.

Furthermore, the use of blockchain technology can significantly improve accountability and transparency within the law enforcement and criminal justice system. By creating a tamper-proof chain of custody, establishing decentralized identity management, improving data sharing and management, authenticating body camera footage, and using smart contracts for automation, blockchain can help reduce errors and corruption, increase efficiency, and build trust between law enforcement organizations and the public. However, it is important to ensure that the implementation of blockchain technology is done with careful consideration and in collaboration with all stakeholders to ensure that it is effective, secure, and ethical.

Decentralized identity management and data sharing are two major challenges faced by law enforcement agencies, not just in the USA, but around the world. Decentralized identity management refers to the ability of individuals to manage their own digital identities without the need for a central authority or intermediary. In the context of law enforcement, this means that individuals should be able to control who has access to their personal information and how that information is used. This can be particularly challenging in a decentralized law enforcement system, where different agencies may have different policies and procedures for managing personal information.

Data sharing and management is another major challenge faced by law enforcement agencies. The ability to share and analyze data in real time can be critical to preventing and solving crimes. However, the sheer volume of data that law enforcement agencies collect can make it difficult to share information effectively. Additionally, the sensitive nature of some law enforcement data, such as information about ongoing investigations or confidential informants, can make it difficult to share that information without compromising the integrity of the investigation or putting individuals at risk. To address these challenges, law enforcement agencies are increasingly turning to technology solutions, such as blockchain and artificial intelligence, to improve identity management and data sharing. These solutions can help ensure that individuals have greater control over their personal information and that data is shared securely and efficiently between agencies. However, implementing these technologies can be challenging and requires a significant investment in resources and training.

It is important to note that the use of blockchain technology is not a panacea for all the issues facing law enforcement. It is just one tool among many that can be used to improve policing practices. Furthermore, the implementation of any new technology must be done carefully and thoughtfully, with a clear understanding of its potential benefits and drawbacks. Finally, it is crucial that any technology solutions used in law enforcement are developed and implemented in a way that respects the privacy and civil liberties of all individuals.

Chapter Summary

There are different classifications and levels of intelligence within law enforcement, and there is a need for more transparent and ethical policing platforms. The use of blockchain-based solutions can potentially address some of these issues by providing a secure and transparent platform for data management and sharing. Blockchain technology can enable law enforcement agencies to maintain an immutable and tamper-proof record of all their activities, including investigations, intelligence gathering, and resource allocation. This can help build trust between law enforcement and the public, as well as increase the efficiency and effectiveness of law enforcement operations.

Immutable records: Blockchain technology provides an immutable and transparent record of all transactions that occur on the network. This capability can be leveraged in the criminal justice system to provide an unalterable record of all actions taken during an investigation or trial. This means that all evidence gathered, all statements made, and all decisions taken can be recorded on the blockchain, providing a clear and transparent record of the entire process. This will reduce the potential for fraud, corruption, or tampering with evidence, and increase trust in the criminal justice system.

Consensus-based decision making: Blockchain technology allows for consensus-based decision-making, wherein parties involved in a transaction or process must agree to a decision before it can be recorded on the blockchain. This means that in the criminal justice system, decisions made by judges, juries, and other stakeholders can be recorded on the blockchain only if they are approved by all parties involved. This can help ensure that decisions are fair and just and reduce the potential for biased or discriminatory decisions.

Smart contracts: Smart contracts are self-executing contracts with the terms of the agreement between buyer and seller being directly written into lines of code. They can be used to automate processes in the criminal justice system, such as the transfer of custody of evidence or the release of prisoners. This can reduce the potential for human error or corruption and ensure that processes are followed correctly and consistently.

Ultimately, the use of blockchain technology in the criminal justice system can increase transparency, reduce the potential for fraud and corruption, and ensure that decisions are fair and just. However, it is important to note that the implementation of such technology must be done carefully and with the involvement of all stakeholders to ensure that it does not inadvertently create new issues or exacerbate existing problems.

Blockchain technology has the potential to revolutionize law enforcement by creating a more efficient, secure, and transparent system for sharing data and communicating between agencies. By using a decentralized and tamper-proof database, blockchain can provide a single source of truth for law enforcement data, thus, reducing the risk of errors, fraud, or data manipulation. One of the main benefits of blockchain technology is that it allows for secure and efficient data sharing between multiple parties. This means that law enforcement agencies can easily share information about criminal activity, suspects, and investigations with each other, without the need for manual data entry or complex data sharing agreements.

Modernizing law enforcement technology is a complex task that requires a multi-faceted approach. It involves addressing the hardware and software components of the system and ensuring that the system is integrated with other systems and technologies used by law enforcement agencies. One way to leapfrog through decades of neglected updates is to implement a cloud-based solution. Cloud technology enables law enforcement agencies to access data and applications from anywhere, at any time, on any device. It can also provide cost savings, scalability, and flexibility, which can be particularly useful for smaller law enforcement agencies with limited resources. An added important consideration is the use of artificial intelligence (AI) and machine learning (ML) technologies to enhance law enforcement operations. AI and ML can help law enforcement agencies analyze large volumes of data, identify patterns, and make more informed decisions. This can lead to better crime prevention, faster response times, and more effective investigations.

In addition, mobile technology can provide officers with real-time access to critical information, such as criminal histories, active warrants, and suspect descriptions. This can help officers make better decisions in the field and increase their effectiveness. It is also important to address cybersecurity concerns when modernizing law enforcement technology. This includes ensuring that systems are secure, and that data is protected from unauthorized access. Therefore, modernizing law enforcement technology requires a comprehensive approach that considers all aspects of the system. By leveraging cloud technology, AI and ML, mobile technology, and cybersecurity measures, law enforcement agencies can leapfrog through decades of neglected updates and improve their operations.

Blockchain technology can also help to improve the security and integrity of law enforcement data. By using cryptographic algorithms and consensus mechanisms, blockchain can ensure that data is not tampered with or altered in any way, reducing the risk of data breaches or other security incidents. While blockchain technology is not a silver bullet for all law enforcement challenges, it has the potential to significantly improve communication and data sharing between agencies, leading to more effective and efficient law enforcement efforts. Different government agencies need to collaborate and share data. This can be challenging due to concerns around data security and privacy, as well as the bureaucratic nature of government agencies. However, there are several ways that agencies can work together to achieve their shared goals:

- **Data sharing agreements:** Government agencies can enter into formal agreements to share data. These agreements can outline the types of data to be shared, how it's shared, and what security protocols are placed to protect the data.
- **Interagency task forces:** Agencies can form task forces that bring together representatives from different agencies to work on a specific issue or problem. These task forces can leverage the expertise and resources of multiple agencies to achieve their goals.
- **Cross-training:** Agencies can provide cross-training opportunities to their staff to help them understand the roles and responsibilities of other agencies. This can help to break down silos and improve communication between agencies.
- **Technology:** Advances in technology can facilitate data sharing and collaboration between agencies. For example, data can be shared securely over encrypted networks, and artificial intelligence tools can be used to analyze large datasets to identify trends and patterns.

By working together, government agencies can achieve their shared goals more effectively and efficiently. This can lead to better outcomes for the public and improved public confidence in government services.

Take Away Lessons from the Chapter

1. Deterring illegal use of digital assets by building a regulatory framework to support digital currencies, regulators can mitigate risks, exploits, crashes, and thefts and provide current regulators like the SEC and CFTC oversight to protect the consumer.
2. The advantages of blockchain technology include expanding access to safe and affordable financial services and reducing the cost of domestic and cross-border funds transfers and payments, transforming the modernization of public payment systems.
3. Illicit finance and national security risks generated by the misuse of digital assets can be mitigated through several measures, including implementing robust Know Your Customer (KYC) and Anti-Money Laundering (AML) measures, regulating digital asset service providers, enhancing international cooperation, increasing public awareness, and developing new technologies.

4. Traditional databases typically use a centralized architecture where data is stored in tables on a server or group of servers, and users access the data through an application or interface. In contrast, blockchain technology uses a decentralized architecture where data is stored in blocks that are linked together in a chronological chain.

5. Each block in a blockchain contains a set of transactions or data records, and when a block is filled up to capacity, it is time-stamped and added to the chain. Once a block is added to the chain, it cannot be modified or deleted, providing an immutable record of all the transactions that have taken place on the network.

6. The immutability characteristic of blockchain technology provides a high degree of transparency and accountability, as anyone with access to the network can see all of the transactions that have taken place. Moreover, the use of cryptography in blockchain technology ensures that the data is secure and cannot be altered or tampered with.

7. Data fragmentation, manual paperwork, and lack of innovation are significant challenges facing law enforcement technology.

Chapter Exercises

1. What are the blockchain platforms that are currently available for use? Why are they effective?

2. Why should organizations evaluate the use of blockchain? What are some of the questions that organizations need to ask prior to implementation of blockchain technology?

3. Explain the main functions of blockchain technology and how it is useful for businesses. Use the framework provided in the chapter to build on the framework.

Blockchain Applications and Uses: Supply Chain Management

Blockchain technology is presently at the center of the supply chain management industry and has successfully provided robust solutions for the industry. The supply chain industry can experience numerous advantages from integrating into blockchain-based networks. One advantage of blockchain is the centralized database server that allows data privacy by generating symmetric keys, utilizing codes and timestamps, and using asymmetric key cryptography for validating and identifying transactions. The supply chain includes several upstream components such as raw materials, suppliers, manufacturing, distribution, and retail/consumer. Embracing blockchain technology for supply chain management creates countless efficiencies and strategic advantages for organizations. Currently, the supply chain management industry has experienced glaring issues brought to light by the COVID-19 pandemic that crushed the supply chain globally. For instance, maintaining a wide variety of inventoried products creates complications. Additionally, product backlog, defects, and friction are caused due to influx of inventory that is hard to manage. Furthermore, tracking products back to suppliers creates challenges when there are defective products. The construct of the supply chain inherently involves many stakeholders. The information collected from one party is not shared with the other party. In this chapter, a comprehensive framework and application of the framework are displayed in detail to provide a roadmap to understanding the use and importance of blockchain-based solutions in the supply chain management industry.

Introduction

Blockchain technology is currently at the heart of goods and services, urban governance, industrial economy, and ecological livability. The financial field has embraced cryptocurrency; however, its underlying technology of blockchains still seems elusive to the industry (Liu et al., 2022). While the advent of blockchain technology has created Bitcoin in the financial sector, a remarkable amount of interest has been percolating among industry leaders to understand and apply blockchain technology in non-financial applications. Mohit et al., (2021) stipulate that the supply chain industry can achieve numerous advantages by integrating with blockchain-based networks. One of the advantages of blockchain is the centralized database server that allows data privacy by generating symmetric keys, utilizing codes and timestamps, and using asymmetric key cryptography for validating and identifying transactions. According to a survey conducted by Deloitte, approximately 79% of organizations that have superior supply chain systems versus only 8% experience above-average growth (Kshetri and Loukoianova, 2019). Therefore, advancements in supply chain systems are a strategic advantage for an organization. The supply chain includes several upstream components such as raw materials, suppliers, manufacturing, distribution, and retail/consumer. Embracing blockchain technology for supply chain management creates countless efficiencies and strategic advantages for organizations.

Weak supply chain management wreaks havoc, as asserted by the recent COVID-19 outbreak, where global trade structures proved to be unsustainable and collapsed. Many items went missing from retail shelves for months because the raw materials needed to manufacture them were not available due to shortages and closures worldwide. For instance, one of the essential items in high demand was disinfecting wipes. The main ingredient that is used in the wipes is heavy paper and bleach. Both raw materials of heavy paper and bleach are manufactured in China and then shipped to the United States to manufacture the product. However, since China and other countries around the world were under lockdown, the raw materials were not shipped to the manufacturing centers to produce the disinfecting wipes. Using this one example, manufacturers of disinfecting wipes could have procured raw materials from within the United States had they maintained records that were easily accessible for local raw materials manufacturers. Blockchain technology creates a decentralized repository of data sharing for all users. If the consumer manufacturing industry collaborates data sharing on a blockchain, many future hindrances in raw material procurements can be avoided.

Many organizations learned valuable lessons through the COVID-19 pandemic. Landi (2020) launched a blockchain network to combat supply chain issues caused by COVID-19. Consumer-driven economies sustain a larger blow from bad supply chain management because consumer-driven economies must carry a wide array of inventory of products that are pleasing to the consumer. Maintaining a wide variety of inventoried products creates complications. For instance, varied inventories create product backlog, defects, and friction. Tracking products back to suppliers creates challenges when there are defective products. The construct of the supply chain inherently involves many stakeholders. The information that is collected from one party is not shared with the other party. Therefore, a simple transaction becomes a very time-consuming and lengthy process that involves several steps. In a traditional supply chain system, only manufacturers add data thus it is hard to track the product back to the original supplier.

Some of the most appealing characteristics of blockchain are speed, transparency, and accessibility. All these features of a blockchain are very attractive as solutions to legacy systems currently managing masses of data that are fragmented. Landi (2020) maintains that a blockchain-based supply chain enhances the privacy of traditional data and implements it in such a manner where traceability and ownership remain intact. In addition, blockchain-based algorithms facilitate the transfer of ownership and trace products through the encrypted ledger. Blockchain provides higher throughput on data security than traditional systems.

A traditional supply chain system has struggled with forgery and fraud. Blockchain offers improvement in supply chain management as it secures data by storing it in a secured repository. Blockchain benefits in the supply chain include ensuring the value of goods reflects in the value of manufacturing costs. According to Kshetri and Loukoianova (2019), Walmart tested the use of blockchain technology in supply chain management to put controls around cost efficiencies. A giant such as Walmart has a colossal source of data, however, has battled many struggles with tainted items that have caused food safety issues. Walmart's willingness to adopt blockchain makes them an industry leader and reinforces its position in the supply chain leadership within the industry.

Customization within blockchain where transactions are encrypted controls the risk of privacy leakage and security attacks. Kshetri and Loukoianova (2019) recommend that to protect data privacy, blockchain must satisfy the linking between transactions that are traceable and transaction data must be encrypted. This means only authorized users can access the data through an encryption key and while centralized systems are good for maintaining privacy, they are not good for product ownership traceability. Blockchain is useful in proving ownership of the product.

Some steps that create a framework for an effective blockchain supply chain are as follows:

1. A manufacturer puts the products on the blockchain.
2. Next, the manufacturers and the distributor transfer the product.
3. The owner of the product is limited to only transfer-based transactions.
4. All parties, manufacturers, distributors, and customers can trace the product.
5. Everyone on the supply chain accesses the same blockchain for information.

Smart Contracts

Smart contracts use symmetric key encryptions. Two types of transactions exist in a smart contract: incoming transactions and stored transactions. Furthermore, a transaction contains four items: hash of product, action, encrypted data, and timestamp. Yavaprabas et al. (2022) performed a systematic literature review to substantiate three dimensions of trust, the trustor-trustee perspective, forms of trust, and time orientation, and discovered that while trust is a naturally embedded factor of blockchain, mainstream literature largely supports that blockchain installation can create trust which subsequently adds trusting relationship between trading parties.

The supply chain has relationship management at its core. The relationship between supply chain partners involves a certain level of risk that creates a need for trust to manage the risks. Trust is a relationship that exists between two entities. In a supply chain relationship, one party (buyer) remains vulnerable because it agrees to carry confidence in the actions of another party (supplier) and has an expectation that the party will fulfill its obligations. Therefore, currently, organizations are seeking to adopt blockchain for supply chain partners based upon trust reinforcement. The next section provides an overview of some major global manufacturing and provides a glimpse into current issues that can potentially be resolved through the adoption of blockchain-based solutions. Six major global industries are included in the discussion—global car and automobile manufacturing, global fruit and vegetable processing, global pharmaceutical and medicine manufacturing, global sugar manufacturing, global cheese manufacturing, and global cosmetics manufacturing. The purpose of including these major global industries is to enable understating of the diverse needs of supply chain management in different industries and actively understand the benefits of blockchain technology as a viable solution.

Global Car and Automobile Manufacturing

According to IBISWorld Report (2022), the global car and automobile manufacturing industry is in for a rocky five years starting in 2022 due to the COVID-19 pandemic slowdown aiding to less consumer, business, and government spending. While the industry is set to hit its business in the next five years it is estimated that global manufacturers will continue to heavily invest in technology and innovation to achieve a wide range of goals to achieve competitive advantage. While the pressures of rapidly growing players in emerging markets are a threat to the industry, a stable supply chain is key to maintaining an overall profit. IBISWorld Report identifies key success factors for the global car and automobile manufacturing industry as:

(a) flexibility in determining expenditure by controlling costs, especially in developed nations to be more competitive;
(b) use of the most efficient work practices that entail close relationships with suppliers and good distribution;
(c) access to the latest available and most efficient technology to enable a competitive edge.

The global auto industry is aggressively embracing technological changes, especially along the automotive supply chain. There is a consistent move towards specialization of the automotive supply chain. The global car and automobile industry are required to comply with government regulations to control emissions, pollution, and safety. Although there are varying requirements in different jurisdictions, global standards are starting to emerge with an increasingly global automotive supply chain. In the US, federal laws mandate that manufacturers recall vehicles if there are defects found that are unreasonable and pose a safety risk. The National Highway Traffic Safety Administration (NHTSA) is the federal agency that compiles data from consumer complaints and commands that unsafe vehicles are recalled. The European Union operates on a similar set of objectives regarding emissions and consumer safety and has analogous testing in place to protect consumers from corporate malfeasance. Japan has implemented emissions and quality control measures and has initiated measures to regulate and reduce emissions. In the same vein, China enacted its first emission control in 2000 and the Chinese government has increased regulations on vehicle pollution.

Global Fruit and Vegetable Processing

The global fruit and vegetable processing industry experiences varying levels of technological complexity due to processing establishments that greatly vary from small-scale producers in underdeveloped countries to sophisticated and advanced processing centers. Developing countries produce most of the fresh fruits and vegetables. However, the product must be shipped to a processing plant usually housed in the developed world. There is a huge opportunity for supply chain improvement for the industry by using technological advancements. According to the IBISWorld (2021) report it is expected that manufacturers in Europe and Asia will produce more than 70% of global exports for the year. There are incredible opportunities for developing a strong supply chain management infrastructure to bridge the gap between where produce is grown and where it is processed. The sheer nature of the business is concerned with spoilage and food waste, and establishing a blockchain technology-based program facilitates data sharing between all parties to ensure there are proper logistics and strict time constraints that are met to avoid spoilage. The Kraft Heinz Company uses an electronic supply chain network to reduce distribution costs and enhance trading relationships between distributors and suppliers. Kraft Heinz's migration of its global data center assets to the new platform is to enable custom workflows, data reporting, and applications to connect to existing supply chain systems. Industry leaders are driving informed actions across the supply chain to gain advantages and minimize constraints. Visibility and flexibility are two main factors necessary to streamline decision-making and improve data visibility and communication across the firm's departments to reduce fragmented efforts. Global food manufacturers must adhere to health and food regulations. Currently, the food industry safety laws are fragmented and food regulations in various countries have different criteria that must be met. Food regulations are aimed at food hygiene and safeguarding consumers against illnesses caused by poor food safety. In the EU, European Food Safety Authority (EFS) which was established in 2002 after several outbreaks and crises, protects consumers and manages risks associated with food supply. Australia has joint effort food standards with New Zealand that are stringent and place much of the burden of reporting and quality control on the manufacturers. In the US, the US Food and Drug Administration (FDA) regulates food safety and while not as rigorous as in Australia, the FDA monitors food safety for consumer benefit.

Global Pharmaceuticals and Medicine Manufacturing

Technological improvements in the global pharmaceuticals industry improve supply chain efficiency and regulatory compliance by lowering operational costs. The regulations in the industry are stringent, numerous government agencies and policies must be followed for manufacturing, pricing, and marketing of the product. According to the IBISWorld Report (2022), an increase in regulations poses the highest potential threat to the industry. Pharmaceutical production typically involves numerous manufacturing components for a single finished product. Therefore, the global pharmaceutical and medicine manufacturing industry must rely on many countries to procure and process one single product. The IBISWorld Report (2022) points out that technology-sharing agreements among major players are a key success factor for the industry. Counterfeiting is a massive problem in the industry. According to the World Health Organization (WHO), an estimated 10% of globally manufactured medication and over one-third of pharmaceuticals in developing countries are counterfeit.

Government regulations pose a significant issue for the pharmaceutical industry as it is one of the most highly regulated industries on a global scale. Fragmented regulatory frameworks that vary by country present another potential barrier for the industry. The high cost of compliance and meeting regulations in various aspects of the production process such as marketing, manufacturing inspections, licensing, and clinical trial regulation adds complexity to the supply chain management of the industry. Furthermore, the high cost of government-mandated testing trials and compliance creates another hindrance to smooth supply chain operations. Globalization in the pharmaceutical industry is on the rise because of significant savings realized by broad geographical distribution and production of products. IBISWorld Report (2022) says that there have been numerous cross-border mergers and acquisitions in the industry to cultivate collaborative alliances across the world, especially considering the COVID-19 (coronavirus) pandemic. Due to the capital-intensive nature of the industry, several large corporations have sought to outscore their manufacturing process to contract manufacturing organizations (CMOs) to aid in improvements in operations and productivity. As such, CMOs has become an integral part of the supply chain, and focusing on specialized niche technologies can produce strong competitive advantages for firms. Information and communications technologies have produced many benefits such as enabling the mechanization and automation of various parts of the drug discovery and development process to help mass production. The supply chain has stimulated changes by bypassing the traditional wholesale to deal directly with consumers instead and downstream technology developments have dramatically increased e-prescribing.

Heavy government regulations in the industry incentivize robust use of supply chain traceability and accountability to preserve and ensure the safety of drugs consumed by the public. In the US, the FDA controls all drug testing, manufacturing, labeling, and approvals. In the EU, the Centralized Procedure (CP) is required by manufacturers of biotechnology and new chemical firms that wish to sell across the EU. In Japan, pharmaceutical manufacturers are required to submit a new drug application to the Pharmaceutical and Medical Devices Agency (PDMA). Clearly, the pharmaceutical industry is heavily regulated by different government bodies that stipulate different regulatory parameters. The utilization of blockchain technology creates an innovative way to solve the fragmented regulations commanded by different jurisdictions by the deployment of smart contracts that especially comply with jurisdictions.

Global Sugar Manufacturing

The global sugar manufacturing industry has experienced massive fluctuations in pricing largely due to increased production from developed nations flooding the market. According to the IBISWorld Report (2022), Brazil has been an important player in the overall health of the industry. Brazil produces and exports more sugar than any other nation on the planet and is currently the second-largest producer of ethanol. Increasing energy prices have lured Brazilian sugar manufacturers to reroute their production for ethanol production, thereby decreasing the supply of raw sugar that causes steep fluctuations in the prices of sugar. The price volatility of the world price of sugar can be monitored through a robust supply chain management system to track the production of sugar in various countries for consumption of raw sugar or routing to earmarks for energy production of ethanol. The industry report identifies supply contracts, operating in a more regulatory-friendly environment, and automation as key success factors of the industry. Creating a robust blockchain-based supply chain platform supports contracts with supplies of raw materials that can be automated. Guaranteed supplies at pre-established prices create efficiencies by minimizing supply costs and planning production. Expansion of downstream activities of the supply chain is advantageous in regions where there are favorable regulations for the industry. The advanced technology of the blockchain structure maximizes output and reduces costs. Government regulations and policies affect the global sugar manufacturing industry by restricting market access for sugar producers from other countries. While the EU and US have revisited sugar trade barriers to reduce market restrictions, there are potential markets where sugar is differentiated by quality, good quality, or superior quality of production. Again, the utilization of blockchain technology to house all data among industry leaders and shared information for government regulations will facilitate competitive advantages for the industry.

Global Cheese Manufacturing

It is projected that global cheese consumption will steadily rise due to relatively strong economic growth and rising affluence in emerging markets. India and China are among the highest consuming nations and cheese is now considered one of the world's largest commodities. Although the industry is predicted to experience steady growth, there are external factors that affect the industry's performance. The quality and price of milk are key inputs to cheese manufacturing. Pandemic-related disruption to the supply chain showed strong downstream demand for cheese products. In many developed nations, the dairy industry is one of the most heavily regulated agricultural markets. For instance, in the US the dairy products market is well above the world market prices; therefore, the domestic US market must be protected against the foreign competition to ensure that the market channel for domestic farmers is stabilized otherwise they would have difficulty selling at higher prices. Some efficiencies that are much desired in the industry are finding ways to reduce waste and energy. As showcased by the Bel Group in 2017 a collaboration with France's Alternative Energies and Atomic Energy Commission (CEA) improved Bel Group's manufacturing processes by reducing waste and energy consumption. The IBISWorld Industry Report outlines a very crucial point: the global cheese manufacturing industry varies significantly across countries and currently there is no global data to provide accuracy of production volumes. This is a key area of focus that can present many strategic opportunities to track products using blockchain technology. One of the key success factors that are extremely important in the industry is product quality.

The use of fresh, high-quality ingredients yields the production of safe and reliable products for the industry thereby maintaining higher profit margins. Constructing strong relationships with upstream and downstream suppliers is critical to being competitive in the industry. Creating a supply chain management system that guarantees access to high-quality inputs to manufacture and secure premium space in supermarkets sets the market leaders apart from their competitors. Moreover, regulatory measures related to food standards and hygiene are extremely stringent in developed countries like Europe and the US. High compliance costs may be too high and pose a potential barrier to growth. Therefore, creating transparency and accountability to establish a higher quality ingredient not only satisfies the most discerning consumer but also creates an avenue for regulators to be satisfied in a cost-efficient manner using blockchain-managed supply chain infrastructure.

Global Cosmetics Manufacturing

According to the industry report, the global cosmetics manufacturing industry is expected to experience strong growth over the next five years to annualize approximately $387.1 billion in revenue (IBISWorld Report, 2021). Revenue growth in the industry is driven by the downstream demand of wholesalers and retailers that are ultimately supported by consumer spending. According to the IBISWorld Report (2021), a key factor to success is the ability to accommodate environmental requirements. There is a steady rise in ethical consumerism that has environmental concerns at its core and is becoming an increasingly important factor for users. Blockchain technology prevails in this area as it provides ethically sourced materials for the end product and assures consumers of the quality and source of the products. Price and quality are two additional critical factors for internal competition in the global cosmetics manufacturing industry. High-quality items in the industry command a premium.

Additionally, several large organizations take advantage of globalization and therefore must steadily maintain their supply chain effectively. At the base of manufacturing the industry involves mixing and blending various ingredients in batch operations. Furthermore, an array of chemicals is used in the manufacturing of cosmetics and personal care products such as fats and oil, fragrances, minerals, cleansing agents, and waxes. As noted earlier, consumers are keen on clean and ethically sourced environmental elements. The global cosmetics and manufacturing industry has huge benefits in utilizing blockchain-based supply chain architecture to incorporate data transparency for consumer demand by providing the source of each raw material used in the end product and connecting this data worldwide on one robust platform. Large organizations in the industry are currently operating in an increasingly expanding globalized environment that is subject to regulations that are related to the ingredients used in the production, packaging, and marketing of the product as specified by each jurisdiction. Thus, incorporating smart contracts within the blockchain that houses specific data on the rules and regulations of a specific country is hugely beneficial in terms of cost savings for the industry.

Structure

The first example of a supply chain can be traced back to Henry Ford, possibly the first supply chain manager to successfully integrate the supply chain in 1926 for Ford manufacturing (van Hoek, 2021). Supply chain management was first introduced in 1982 in the Financial Times interview by a Booz Allen Hamilton consultant (Berthold, 2019). As globalization set in, mass production and advancements in information technology in the 1990s catapulted implementation of supply chain management practices in Fortune 500 companies all around the world (Min et al., 2019). The adoption of supply chain management proved to be a critical contribution to organizational success. Thakkar et al. (2011) highlight the many benefits of supply chain management, reduced disruptions, improved product quality, shorter lead times, reliable logistics, and cost savings.

While organizations realized the importance of supply chain management, the vast and complex nature of the process created many deficiencies that remain unresolved. The evolution of global business has created an increasingly complex, interconnected, and volatile business environment that increases vulnerabilities and makes it more critical than ever for supply chain management to be flexible, resilient, and adaptable (Alfarsi et al., 2019; Ambulkar et al., 2016; Nguyen et al., 2019). Among the many definitions of supply chain learning Yang et al. view supply chain learning as a process, structure, or consequence orientation (Yang et al., 2019). Process orientation of supply chain learning addresses the capacity to build on new knowledge and improve the supply chain process through inter-organizational and collaborative knowledge construction (Flint et al., 2008; Lambrechts et al., 2012). Structure orientation is the supply chain learning behaviors in an organizational context as compared to the establishment of new techniques (Bessant et al., 2003; Ojha et al., 2018b). Consequence orientation is an organization's willingness to encourage organizations integration and developing new insights that impact supply chain capabilities and intangible resources that drive supply chain success toward a competitive advantage (Biotto et al., 2012; Ngai et al., 2011; Sweeney et al., 2005).

Learning is a key component of supply chain management. The supply chain learning process includes a set of procedures to promote supply chain learning, operating, and governing to continuously sustain knowledge through the management of processes and measuring benchmarks (Jia et al., 2019). Learning is a pivotal component of supply chain competency, which promotes reducing uncertainty in supply chain decision-making through continual collaborative learning and knowledge sharing (Asmussen et al., 2018; Gong et al., 2018). Information sharing among organizations builds knowledge, skills, competitive advantage, and intellectual capital (Ataseven et al., 2018). Sharing information enables supply chain management functions to leverage supply chain integration, strategic alignment, and developing collective interactive capabilities (Butt, 2021). Rebolledo and Nollet (2011) believe that inter-organizational hurdles are a lack of auxiliary structures, procedures, and trust. Dynamic organizations are constantly seeking innovation to strengthen their supply chain. Overstreet et al. (2019) cite technology barriers as a main shortfall in supply chain management. Organizations are constantly evolving and adapting to successfully integrate and strengthen their supply chain management system (Flothmann et al., 2018). Many authors agree that building a resilient supply chain system required building, categorizing, and storage of data inter-departmentally to leverage collaboration, supplier management, and strategy development (del Rosario Pérez-Salazar et al., 2017; Schniederjans et al., 2020; Wahab et al., 2021).

Several authors assert that supply chains have become increasingly difficult to manage due to the progression and complexity of global inflows and outflows presenting challenges for organizational learning capabilities (Bals et al., 2019; Derwik and Hellstrom, 2017; Haq, 2020; Hult et al., 2000; Jordan and Bak, 2016; Mageto and Luke, 2020; Ojha et al., 2018a; Overstreet et al., 2019; Tatham et al., 2017; Zhu et al., 2018). The evolution of supply chain management has created enterprise complexities due to manual processes and human interactions. Mello et al. (2021) believed that human interactions in supply chain management required alignment of the right individuals performing the right tasks to create maximum efficiency. Traditionally supply chain management functions are imparted by unwanted behaviors that impact inventory management and therefore affect organizational efficiency and profits (Cormack et al., 2021; Derwik and Hellstrom, 2021; Jia et al., 2019; Selviaridis and Spring, 2018; Wiewiora et al., 2019). Supply chain learning is often obscure because of its complexity, delayed feedback, manual processes, heavy human interactions, behavioral factors, and organizational learning culture (Bessant et al., 2003; Tokar et al., 2012; Wiewiora et al., 2019). In addition, delayed information collection, limited information flow and sharing, asymmetrical data sharing, and common trust issues create oscillation in operations that amplify instabilities across the supply chain (Sterman et al., 2015). Therefore, supply chain decision-making can be optimized by identifying alternative methods of counteracting unwanted behaviors to improve supply chain management and learning decisions.

What are some of the factors to consider elevating traditional supply chain methods? Addressing the human element, creating a dynamic system of collaboration, responding to organizational success and competitiveness, and increasing efficiency all lie at the heart of improvements surrounding traditional supply chain management systems. Attia and Eldin (2018) point out that competencies in supply chains have significantly transformed over the years and believe that organizations must constantly improve their learning ability to advance their probabilities of long-term success. Human decision-making is a strong element that affects the improvement of supply chain management processes and decision-making (Fahimnia et al., 2019). Increased global trends and competition demand organizations to leverage leadership to increase collaboration among supply chain partners (Heyler and Martin, 2018). Organizations must understand that supply chain managers should consider understanding the cause and effect of the relationship between employee decisions and behaviors to improve decision-making skills (Haines et al., 2017; Shou et al., 2018). One of the most significant factors that increase complexity in supply chain management is human interaction (Singh et al., 2021). Optimized supply chain performance is no longer dependent upon hierarchy or ownership, instead, important for interactions between the supply chain associates (Goudarzi et al., 2021; Patel, 2017). Many authors have asserted that individuals lack the capacity to act rationally while decision-making (Mageto and Luke, 2020; Schorsch et al., 2017). Organizational leaders must understand that there is a direct correlation between supply chain decision-making and individual behaviors that impact inventory management and far reaches all aspects of the supply chain. Leaders can gain substantial value by removing individual differences among decision-makers as it could have a significant bearing on supply chain management (Hofstra et al., 2019; Li and Yan, 2015). Creating value in the inventory management process can produce strategic efficiencies for organizations.

Elmountasser (2019), provides insight into inventory control logarithms that did not account for human factors even though the premise relies on human intervention to correct inherent inadequacies. Inventory management is a complex process that relies upon procuring a stock of material and seeks to determine how much to order, when to order, and how much safety stock must be held against stockouts (Mondal et al., 2020). Bak et al. (2019) state that decision-making skill sets are among the most important in the supply chain management industry. Decisions such as ordering decisions, supplier selections, and purchasing decisions are all made through bounded reasoning that is shaped by contextual factors like the task's difficulty, importance, and compensation (Kull et al., 2014).

Interactive preferences such as overconfidence considerably affect decision-making and negatively impact the organization's cost savings. Individual risk profiles also contributed to purchasing decisions of the overall supply (Zhijian et al., 2020). A person's emotions and illogical behaviors affect their buying decisions causing inadequacies, increased costs, and results in disruptions in the supply chain. The human interface has delivered many inventory complications. One such well-documented phenomenon is the bullwhip effect. Perera et al. (2020) describe the bullwhip effect as an issue caused by demand forecasting, rationing, shortage gaming, order batching, and price fluctuation. The bullwhip effect occurs when small fluctuations in demand cause progressively larger fluctuations in demand and signal to the other participants in the supply chain a potentially higher level of false demand. As decision-makers ignore time interruptions and inadequately contemplate the supply line it leads to the bullwhip effect (Sterman, 1989). Another supply chain event is the newsvendor model that has been studied since the 1950s.

The concept of the newsvendor model is directly correlated with effective inventory management. At the core of the model, it is assumed how many newspapers a news vendor should buy from the printer that day, knowing the unsold copies will be of no value. Therefore, in this model, the organization decides on ordering quantities without knowing price and demand data. The purpose of the newsvendor model is archetypally to study and explore rational inventory pricing, and ordering decisions based upon the assumption that the decision-makers exhibit stable preferences and rational decision-making principles (Sterman, 1989; Ying and Gui-hang, 2021). A common misconception underlying inventory management considers that decisions are made through procedural compliance by rational individuals; however, research has established that individuals make decisions based on their biases and beliefs more often than their training and experience. Despite heavily documented and discussed literature regarding the bullwhip effect and its causes, there has been no efficacious mitigation technique discovered as of yet (Narayanan and Ishfaq, 2022).

Currently, SCM systems must rely upon manual, human decision-making. Teniwut and Hasyim (2020) assert that decision support systems in supply chain management are complex and affect decisions like inventory forecasting and reordering decisions. Decision support systems comprise data analytics, game theory, and forecasting (Viet et al., 2018). Therefore, sharing of data is a critical component that feeds into decision support systems. Many current methodologies require advanced planning and utilize computerized systems that account for constraints and objectives (Allaoui et al., 2019). Application of the enterprise resource systems (ERP) has provided organizational software systems that created integrated modules for procurement, inventory management, material requirement planning, and more; however, ERP systems depend on the accuracy of information entered by a manual process by humans with little to no oversight for data integrity (Brauner et al., 2019). Existing technologies depend upon accountability to achieve supply chain management functions. Accountability holds decision-makers responsible for the quality of the actions used to decide (Garg et al., 2017). The industry's present methods have created an inability to integrate supply chain management decisions and the lack of interventional methods has influenced unwanted behaviors that impact supply management functions negatively.

Framework

Supply Chain Framework

According to a solution brief crafted by IBM, the supply chain is identified to be at the epicenter of providing consumers with products. Regardless of what the organization produces, from grains to computer chips, a reliable and trusted supply chain is the key to customer satisfaction and meeting financial targets (see Figure 8). Building an intelligent supply chain requires seamless integration and coordination among the various stakeholders in the supply chain. While organizations realize the importance of a robust supply chain, currently organizations are facing challenges that have created disruptions due to increased cost that causes loss of revenue. In addition, inefficient processes

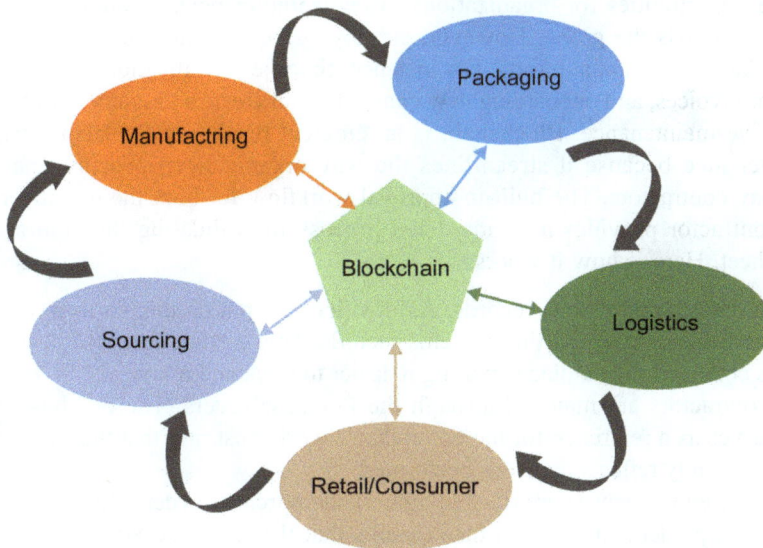

Figure 8: Framework Supply Chain Management Industry
(Source: Author's own work)

that rely upon data that is not trusted and is not available in a timely manner exacerbate these disruptions. Organizations understand that they need to find solutions that can provide transparency, reliance, and agility. Blockchain-based supply chain technology provides trust, transparency, and consensus for all stakeholders. Blockchain technology can be customized to infuse supply chain networks from full production and distribution.

According to IBM Solution Brief, 40% of enterprise leaders stated that data sharing and integrity were the major challenges in their supply chain. In addition, 70% of enterprise leaders upheld that removing human intervention improved the quality of data, speed, integrity, and visibility (IBM, 2022a). The brief shared that IBM has partnered with hundreds of organizations to engage in developing a robust supply chain spectrum to gain speed, save time and create value through blockchain technology. The fabric of the blockchain can be customized for organizations by realizing where significant business value exists so a comprehensive roadmap for future capabilities can be designed.

Disputes occur during transactions concerning two parties in the supply chain and cost a significant amount of time and labor to resolve. IBM Solution Brief (2022b) states that firms hold an average of USD 200 million in disputes, therefore there is an immense opportunity for improvement by efficiently resolving multi-party disputes. For instance, just a three percent improvement in supply chain operations can yield cost savings of upwards of USD six million. Blockchain platform has built in a consensus layer where disputes are settled based upon agreed rules that ensure accuracy, fairness, and time efficiencies for all participants. The advantages of operating in a continuously globalized world create opportunities for organizations to realize efficiencies through contingent workers across the globe. However, employing contingent workers across the globe increases manual efforts. For instance, timesheet verification, validation of vendor invoices, and onboarding new contractors all create a vast amount of labor-intensive maintenance. Blockchain is an efficient resolution for labor-intensive maintenance because it streamlines the way organizations procure, manage, and pay contractors. The built-in approval workflow for both the employer and the contractor provides a standardized process for validating the contractor's timesheet. Here is how it works:

1. Purchase orders, time records, and invoices are put on the platform through the utilization of existing systems records. This ensures that all information is collected in one place, making it easier to track and manage.
2. Contractors are matched through the Purchase orders. The purchase order serves as a reference for the contractor's work, ensuring that their timesheet accurately reflects the work performed.
3. Time and expenses are tracked against the purchase order. This mechanism has a predefined function that ensures that the time tracked by the system matches the purchase order to fulfill this criterion. This means that any discrepancies in the timesheet can be easily identified and resolved.

4. Once consensus is reached, transactions are recorded on the blockchain ledger. This provides a transparent and immutable record of all transactions, ensuring that there is no dispute or confusion over the work performed and the payments made.
5. This is followed by invoice generation, reconciliation, and payment. With all information collected in one place, the process of generating invoices, reconciling them against the purchase order and timesheet, and making payments becomes much simpler and more streamlined. The built-in approval workflow ensures that all parties are satisfied with the work performed and the payments made.

The most challenging part of building a blockchain network for a business is the organization itself. To successfully build a supply chain for an organization's ecosystem, organizations must understand and design a path to their success.

Application

Xie et al. (2022) discovered that blockchain-based databases provide detailed design of information storage, traceability, and smart contract for agricultural products and support the supply chain through dynamic data storage technology. Currently, the business of purchasing goods and services is transacted through the management and supervision of a third party. User information, including transaction information, is stored on third-party servers. The convenience of e-transactions and a lack of traditional physical transactions has created explosive growth for e-commerce. Global industries are constantly seeking quality assurance. For instance, the modern food supply chain has become increasingly complex. The cultivation of food, transportation, distribution, retail, and standardized management of production processes creates more challenges in supply chain management. For instance, the production of food and processing entails a large amount of data that encompasses many participants. An outbreak of diseases, subpar products, and the proliferation of counterfeit products have altogether exposed serious complications regarding transparency in food production and distribution. Blockchain technology offers information traceability where the source of food production and distribution channels are easily identifiable. This resolves issues with pursuing accountability as highly accurate and traceable information is readily available. If promptly identified, the causes of contamination during an outbreak can be effectively acknowledged and mitigated. Dependable traceability of information provides evidence of whether the requirements are strictly met during the distribution process and confirms the healthiness of the food. Price justification and food fraud issues can be minimized through the usage of food information that validates the authenticity of food products and confirms quality parameters.

The rapid evolution of the services has created an ever-increasing transaction volume (Xie et al., 2022). While transactions are growing phenomenally, the storage of the transaction data exists in a fragmented environment where consumers cannot easily trace the required. In addition, organizations store private data on the consumers and are unable to build secured data that remains vulnerable due to data breaches and possible internal data tampering.

Xi et al. (2022) suggested blockchain technology to store food information traceability and proposed food information traceability, scheme, and design for an efficient consensus algorithm for improving transaction rates that is effective. Xi et al. (2022) also maintained that a blockchain-based database precisely allows the problem in traditional agricultural traceability systems to be resolved due to the inherent nature of a distributed ledger technology that lies at the foundation of the blockchain.

Features of the blockchain that create value in information traceability are achieved through timestamps, consensus algorithms, data encryption, and smart contracts which are all components of a blockchain. In addition, the consensus mechanism built in a blockchain application creates a tamper-resistant environment that alleviates fraud and information symmetry during agricultural product information collection and transaction. A centralized approach to data sharing achieves full supply chain traceability and allows the right of entry to anyone who has access to the database. Traceability of food production creates value for all stakeholders in the flow of agricultural products and provides visibility to consumers of real information about the agricultural products they are purchasing. Chang et al. (2020) research provided insight into the consumer electronics industry and the authors maintain that blockchain technology provides the consumer electronics industry with more transparent, secure, and honest transaction activity. Furthermore, blockchain transactions and regulations can be managed through smart contracts and proposed a regulation verification approach by incorporating smart contracts.

Blockchain features authenticity of the existence of transactions through an algorithm and timestamp. The work uploaded to a blockchain is verified and the following transactions are recorded in real-time. To effectively track food information, the traceability of food production needs to achieve end-to-end traceability. The nodes on the blockchain need to be updated with food transfer information to avoid breakage and ensure the relevance of traceability information to cover the entire process of food circulation. In practical application, after the food information transaction and other pertinent supply chain transactions concerning the food are uploaded to a blockchain database, the data is opened to the entire network that consists of all parties involved including regulators who have access to swiftly locate and investigate offenders.

Utilizing blockchain technology to maintain relevant information on agricultural products creates consumer confidence and strategic advantages for participating businesses. Blockchain benefits include improved standardization, consciousness, and reliability that reduce the occurrence of infringement and piracy. Agi and Jha (2022) claim that even though blockchain technology has garnered significant importance and acceptability, there are missing factors that are critical to properly adopting blockchain technology in the supply chain. Agi and Jha (2022) discovered 20 enablers of blockchain technology adoption for the supply chain and assert that blockchain technology is a promising technology that applies to supply chain management.

Although blockchain-based technology provides a powerhouse potential, its adoption in supply chain management is underexplored. Blockchain technology provides a platform for peer-to-peer network technology that utilizes distributed ledger to create and maintain a database of records. All parties participating in the blockchain network could interact with each other and create records. Prior to records being stored on a blockchain, the records are verified and validated through a consensus mechanism and once validated the records are combined to form a block of data that is linked to each other to form a chain of blocks or a "blockchain". A blockchain-based database is chronological in order, every block on the blockchain contains the hash of the previous block. In addition to the structured database, the entire database is replicated and stored on diverse nodes of the system. A private or a permissioned blockchain is different from a public or not permissioned blockchain because the public blockchain allows everyone on the blockchain access to the data while a private blockchain only allows access to predefined users. Regardless of public or private blockchain structure, blockchain technology is characterized by the implementation of consensus mechanisms that are at the core of validating data using cryptographic links between the blocks of the chain.

In supply chain management there are upstream and downstream flows, for instance, upstream flows include raw material procurement, suppliers, and manufacturing, while downstream include part of manufacturing, distribution, and consumer. The key to successful supply chain management is the ability of all stakeholders to cooperate and mutually share information. Blockchain technology significantly improves supply chain management by providing a platform for direct interfaces between supply chain members to exchange data safely (Afzal et al., 2020; Wamba et al., 2020). Enhancing traceability and visibility through the upstream and downstream flows of the supply chain creates trust. At the core of blockchain technology, the platform enables product traceability and visibility through the different supply chain stages. Retail giant Walmart collaborated with IBM to digitally trace pork products in China from farm to table. Details such as the farm where the pork was produced, the factory where the pork was processed, the batch numbers, the storage temperature of the product, and shipping details were all part of the data stored on the blockchain (Yiannas, 2017). Digital access to the data powerhouse creates visibility of the product from every stage of the supply chain; thereby, giving everyone involved in the blockchain access to information. Many food safety issues continue to occur largely because it takes time to track down the supplier or the issue that contaminated the food. As exhibited by the example of Walmart, all data points are accessible including the temperature of the product storage that could have caused food safety issues. Blockchain-based solutions construct and manage distinctly identifiable data for all users on the chain which provides organizations with an extensive strategic advantage.

Operational efficiency is another advantage of the blockchain technology application in supply chain management. The incorporation of smart contracts builds automatically executable interactions between blockchain users. For instance, if the grain is being procured from Farmer A, once the digital contract is fulfilled, Farmer A has satisfied their delivery of the contract. The blockchain can then record this data to execute payment for Farmer A through the financial institution. In a traditional setting where blockchain technology is missing, typically there is a manual labor-intensive process that must be utilized to record the same data. Many intermediaries are involved in the process. The receiving paperwork generated through the receiving warehouse is sent to the accounting department, where accounting personnel match the purchase order to the receiving paperwork and validates the transaction. Once the accounting department clerk validates the transaction, it is then entered into the system through accounts payable. Accounts payable manually enters the invoice for payment on accounting software and then the payment is queued to be paid per the net terms of the contract.

Once the net terms of the contract are met, an accounts payable clerk generally runs a range report for all invoices that are due in that time frame. After the invoice for Farmer A is identified to be paid, the accounts payable department pushes the payment for approval to the treasury department which validates the payment for approval to be paid through the financial institution. A manager then approves the final payment to be processed through the bank. Every step of the traditional process is heavily burdened with manual work that is prone to human error. Blockchain technology provides a sustainable environment to share data from the different processing to transporting creating operational efficiency for supply chain members. In this example, many manual steps are removed, and the data is validated through the execution of a smart contract to simply pay Farmer A, thereby producing a high degree of operational efficiency. Food safety is mandated by food agencies that regulate food production and distribution. Blockchain technology provides a platform where information is interconnected, and the previous set of data is linked to the compliance and certification requirements of the product to meet the regulatory requirements.

The use of blockchain technology not only increases trust and operational efficiency but also provides a forum for lowering transaction costs for the supply chain participants. The author points out that although blockchain technology has the potential to revolutionize supply chain management, the adoption of blockchain in global supply chains has complexity and rules that must be standardized for interoperability between blockchains.

Understanding Blockchain Technology

Blockchain is a robust technology that was created and brought to the forefront of the technological showcase due to its ability to publicly validate, record, and distribute transactions on an encrypted decentralized ledger. The essence of discovering blockchain technology is utilizing a digital, decentralized platform to create a distributed ledger for recording each transaction of the bitcoin, a digital cryptocurrency whose purpose is to operate independently from a central bank. Since the transactions and ledgers on a blockchain are encrypted, blockchain technology offers elevated security for data storage. For instance, in the banking industry, currently, interest-based programs are utilized to conduct business. It can take two-to-three days to clear transactions and money exchange. The application of blockchain technology in the banking industry eliminates lengthy transactions and turns them into real-time transactions as soon as they are recorded on the blockchain.

Blockchain technology is not only limited to the banking industry but also has potential applications in various other industries such as healthcare, supply chain management, real estate, and more. One of the unique features of blockchain technology is that it allows for trustless transactions, meaning that transactions can be verified and processed without the need for intermediaries such as banks or other financial institutions. This can potentially reduce transaction costs, increase efficiency, and improve transparency.

Another important aspect of blockchain technology is its decentralization, which means that the ledger is not controlled by a single entity or organization but is distributed across a network of computers. This enhances security and reduces the risk of fraud or hacking as there is no single point of failure. In addition to its practical applications, blockchain technology has also generated significant interest in the field of cryptocurrencies, with numerous alternative cryptocurrencies (altcoins) being developed using blockchain technology as the underlying platform. Therefore, blockchain technology has the potential to revolutionize the way transactions are conducted and recorded, offering increased security, efficiency, and transparency in various industries.

Blockchain receives its name from its structure. There are a set of blocks that contain data, which is validated and secured through encryption creating immutable transactions. Once a block is filled with data, a new block is formed and each block on the blockchain is linked together in chronological order. The term "blockchain" refers to the structure of the database that stores transaction data in a secure and decentralized manner. Each block contains a set of transactions that are verified by nodes in the network and then added to the blockchain. Once a block is added to the blockchain, its data cannot be modified or deleted, which creates an immutable record of all transactions on the network.

The blocks in a blockchain are linked together in a chain-like structure, with each block containing a reference to the previous block in the chain. This creates a chronological order of transactions, with each block representing a new set of transactions that have occurred since the previous block was added.

The encryption and decentralization of the blockchain also helps to ensure the security of the network. The transactions on the blockchain are encrypted using advanced cryptographic algorithms, which makes it extremely difficult for anyone to tamper with the data on the network. And because the blockchain is decentralized, there is no central authority that controls the network, which makes it more resistant to attacks and hacking attempts.

A transaction on a blockchain is created when two parties initiate a transaction. Once the transaction is initiated the blockchain automatically assigns encryption. Then the blockchain verifies the transaction data and validates it, creates a block, and stores the data regarding the transaction on the block. New blocks are added when all transactions recorded in one block fill that block. Once all these steps are successfully executed, the blockchain transaction is considered complete and the blockchain ledger is updated.

While the adoption of blockchain technology has not been steadfast, organizations that realize the true value of this technology have reported success in many cases.

- Walmart tested blockchain applications when the company took the initiative to trace pork that was produced in the US and sold in China for authenticating transactions, increasing the accuracy of data and record keeping.
- Maersk, a global logistics and supply chain leader in the shipping industry, has been working on cross-border transactions recorded on blockchain technology to improve process efficacies.
- BHP is one of the world's top producers of commodities, appreciates the robust solutions of blockchain technology, and has implemented the technology to replace spreadsheets and legacy programs to track internal and external transaction data to improve efficiencies.

In the realm of supply chain management, blockchain technology adds value in three major areas.

1. **Replacement of legacy technology, manual processes, and consolidation:** Blockchain application is a large-scale technology that can facilitate massive amounts of data, structure, and streamline business processes; however, lower supply tiers are slow to adopt the technology due to reliance on old manual processes. For instance, the shipping industry predominantly relies upon paper transactions. A change to new technology that revamps how business has always been conducted is not always easy to implement because it requires a change in the industry and participants' willingness to accept, learn, and get trained to learn the new technology.
2. **Reinforcing traceability:** Simplification and agility are two valuable concepts that create value in a supply chain. For instance, in the food industry, there is a movement for consumers demanding to know where the food is sourced. In addition to consumer demand, there is increasing regulatory compliance that needs to be met. Incorporating a blockchain to manage the food supply chain adds value by alleviating the high cost of quality problems that give birth to recalls and creating trust to mitigate reputational damage and loss of revenue.
3. **Reduction of transaction costs:** The volume of data created in a supply chain creates challenges for organizations. For instance, in the automobile

industry, a manufacturer can send billions of blanket purchase orders that allow multiple shipments in a short time. The storage of data of this magnitude raises demand for storage, this is an essential component of the blockchain distributed ledger function.

While there are several advantages to the application of blockchain technology in the supply chain, the adoption of the technology does not eliminate the need for privacy. Participants in a supply chain are faced with decisions on whom to share the information with. The supply chain is unlikely to accept open access because not all users on the blockchain are eager to divulge proprietary information. Thus, important data such as demand, capacities, orders, pricing, and margins are all data points that users may want to restrict from unknown participants. The organizations need to collaboratively decide upon the standardization of blockchain, so all participants are aware of data exchange parameters. A private blockchain can accommodate these criteria as organizations onboard blockchain technology.

While the application of blockchain technology in the supply chain has several advantages, such as increased transparency, immutability, and traceability, it is essential to consider the privacy of the data shared within the blockchain. The supply chain involves multiple stakeholders, such as suppliers, manufacturers, distributors, retailers, and customers, each with its own set of data that needs to be protected. Some of this information may be confidential and sharing it with unknown participants could result in significant risks, such as the loss of intellectual property, reputational damage, and financial losses. In such cases, a private blockchain can be an effective solution that provides the necessary security and privacy for the data shared within the supply chain. With a private blockchain, organizations can define who can access and view specific data, and they can restrict access to unknown participants. This allows organizations to protect their proprietary information while still benefiting from the advantages of blockchain technology.

Furthermore, standardization is essential for ensuring that all participants in the supply chain are aware of the data exchange parameters. Collaborative decision-making on the standardization of blockchain can ensure that all stakeholders have a clear understanding of the rules governing the blockchain. This can help build trust between participants and ensure that the blockchain operates smoothly, with minimal disputes or errors. While the adoption of blockchain technology can significantly improve the efficiency and transparency of the supply chain, privacy, and data protection remain essential considerations. A private blockchain can be an effective solution to address these concerns while still providing the benefits of blockchain technology. Collaborative decision-making on the standardization of blockchain can help ensure that all participants in the supply chain are aware of the data exchange parameters and can build trust and confidence in the blockchain's operation.

The COVID-19 panacea tested the limits of preexisting supply chain networks and revealed that there are serious concerns about the preexisting system. The current supply chain systems are often siloed, fragmented, difficult to access, and hard to analyze in terms of big data. Blockchain technology offers a medium that can revolutionize the industry by creating a set of globally accepted standards. There is a huge untapped potential for industry leaders to create a new wave of technology that creates standardization and moves business processes in the most efficient manner yet known to man.

Alfarasi (2020) claims that blockchain technology is one of the most promising solutions for supply chain management and is supported by several benefits that blockchain offers. Firstly, blockchain technology allows for faster and more cost-efficient delivery of products, which can significantly improve the overall efficiency of the supply chain. Secondly, blockchain enables product traceability, which is crucial for industries such as food and pharmaceuticals, where consumers require detailed information about the origin and quality of products. Thirdly, blockchain technology facilitates coordination between partners by creating a secure, transparent, and decentralized platform for sharing information and executing transactions.

Furthermore, blockchain technology enables a limitless number of anonymous parties to transact privately and securely with one another, which can expand the scope of the supply chain and increase access to new markets. This contrasts with the current construction of the supply chain, which allows a limited number of known parties to interact with each other. By providing an infrastructure that allows scalability to today's supply chain business model, blockchain technology can help improve the overall efficiency, transparency, and security of the supply chain. Thus, the new construction of blockchain technology in the supply chain space requires permissioned blockchains with new standards to represent transactions on the blockchain. Additionally, new governance rules will need to be developed to manage authority and meet regulatory requirements. Therefore, while blockchain technology offers several benefits for supply chain management, it also requires careful consideration and planning to ensure its effective implementation.

Due to the inherent nature of the current supply chain management structure, access to the system is limited because of privacy concerns and a profound desire to not share pertinent organizational data with all users. Traditionally supply chains are not built to share information. However, severe challenges in traditional supply chains have caused many issues for organizations. Blockchain technology provides a futuristic approach to starting fresh with a technology that can resolve most serious pre-existing problems with traditional supply chains. As the industry evolves with blockchain technology there are some key system improvements that the industry must agree upon to provide more decentralized sharing of data with all on the blockchain.

Traditional supply chain management structures are built on a model of limited access to information due to privacy concerns and a desire to protect sensitive organizational data. However, as you also note, this approach has caused significant challenges for organizations. Blockchain technology offers a promising solution to these challenges by providing a decentralized and secure way to share information among supply chain stakeholders. Blockchain technology can enable the secure and transparent sharing of information in real-time, which can lead to increased efficiency, reduced costs, and improved customer satisfaction.

To fully realize the potential of blockchain technology in supply chain management, there are some key system improvements that the industry must agree upon. These include standardization of data formats, protocols for data sharing and access, and governance mechanisms for ensuring the integrity of the blockchain network. Standardization of data formats is essential for ensuring interoperability between different supply chain systems and enabling seamless data sharing. Protocols for data sharing and access can help ensure that only authorized parties have access to sensitive data while still allowing for the transparent sharing of information.

Governance mechanisms, such as consensus algorithms and smart contracts, can help ensure the integrity of the blockchain network and prevent fraudulent or malicious activity (Rymarczyk, 2022). Therefore, blockchain technology provides a futuristic approach to supply chain management that can resolve many of the pre-existing problems with traditional supply chains. However, to fully realize the potential of blockchain technology, the industry must agree on key system improvements to enable more decentralized sharing of data with all on the blockchain.

There are several limitations to how business is currently conducted in the supply chain industry. To explain this further we utilize a hypothetical example, in a simple transaction on the supply chain a retailer will source products from a manufacturer and then pay the manufacturer using working capital as the order fills. This simple transaction requires flows of information, inventory, and finance.

While the supply chain industry is vital to the success of businesses worldwide, there are still several limitations to how it operates. Highlights of the challenges faced by supply chain stakeholders, including retailers and manufacturers, include lack of transparency, inefficiency of payments, and the lack of ability to track inputs. Firstly, one of the main limitations in the supply chain industry is the lack of transparency and visibility across the entire supply chain. For example, the retailer relies on the manufacturer to provide accurate information on inventory levels and delivery times. However, the manufacturer may not have complete visibility into the supply chain, including its suppliers and their capabilities. This lack of transparency can lead to delays, quality issues, and additional costs.

Another limitation is the inefficiency of traditional payment processes. The retailer pays the manufacturer using working capital as the order fills. However, this process can be time-consuming, particularly if there are delays or disputes during the transaction. Additionally, traditional payment processes can be costly, particularly for small- and medium-sized businesses that may not have access to the same financing options as larger companies. Finally, there is the issue of inventory management. In addition, the retailer relies on the manufacturer to maintain accurate inventory levels. However, if the manufacturer experiences production delays or quality issues, this can impact the retailer's ability to fulfill orders and meet customer demand. Ultimately, while the supply chain industry is essential to business operations, there are still several limitations that must be addressed to improve efficiency and profitability. These limitations include a lack of transparency, inefficient payment processes, and inventory management challenges. By addressing these issues, supply chain stakeholders can better collaborate, reduce costs, and improve customer satisfaction.

A traditional supply chain system must perform manual audits and inspections that are not reliable and more importantly, do not connect the three flows, making it hard to eliminate execution mistakes, improve decision-making and resolve supply chain conflicts. However, if blockchain-based technology is utilized, blind spots can be eliminated to construct the relevant flows of information, inventory, and money that are reliable and shared among all three participants. Using the same hypothetical illustration, we designed the steps for capturing details of simple transactions based on conventional supply chain management systems vs. blockchain systems.

1. The retailer places an order with the supplier. The supplier acknowledges the receipt of the order.
2. The supplier communicates with the bank to provide financing for the order. Bank approves and provides financing for the transaction.
3. The supplier ships the merchandise to the retailer and invoices.
4. The retailer pays the supplier.
5. The supplier pays the bank loan and closes the record.
6. The retailer returns damaged and unsold merchandise back to the supplier and the supplier pays the retailer back.

All the above six transactions show how blockchain technology would record the transactions. The key advantage of blockchain is to provide real-time data. This element of the blockchain is incredibly valuable because mistakes in inventory data, missing shipments, and duplicate payments are impossible to track in real time on the conventional supply chain platform. While conventional systems that rely upon Enterprise Resource Planning (ERP) systems capture the data, they cannot provide access to all data that is stored within the system to logically remove execution errors. Especially organizations that are engaged in a large network have a multitude of transactions that are not easy to track on the ERP convention database. Recording and effectively sharing massive amounts of data on the supply chain is just one factor that conventional systems are unable to handle. The supply chain components are very complex. For instance, orders that are not straightforward create more data that may not sync with the corresponding shipments. Multiple shipments may also create more data points that are not captured efficiently.

The auditing function plays a critical role in supply chain execution. The transactions in a supply chain must be verified through audits. Audits lay the groundwork for ensuring compliance; however, even auditing transactions does not improve the quality of decision-making to create operational efficiencies. In the food industry, there is a lot of waste that is created due to expiration. There is no efficient method to track why there is so much food spoilage and what can be done about it. Glitches in the traditional supply chain system, along with weak and sporadic demand for goods, and inefficient inventory management create an environment to increase food waste.

Another area of concern in the traditional supply chain is the construction and validation of records to produce efficient inventory management that can assist in lessening food waste. While ERP systems can assist at a limited capacity, these systems are generally expensive and time-consuming. Incorporating record-keeping through the adoption of blockchain technology could create structured approaches to data maintenance, assign unique identifiers for data security, and provide participants with these unique identifiers for digital signatures to maintain the integrity of the blockchain. Regardless of the complexity of the supply chain activities, blockchain can accommodate all transactions related to record-keeping as inventory, orders, loans, bills of lading, etc., using smart contracts and a methodical and functional manner.

Incorporating blockchain technology in supply chain record-keeping can potentially address some of the challenges faced in traditional supply chain management. By creating a secure and transparent ledger of all transactions and data related to the supply chain, blockchain can help reduce errors and fraud, improve inventory management, and ultimately reduce food waste. Smart contracts can also be implemented to automate certain processes and enforce rules and agreements between parties. For example, a smart contract can be programmed to automatically trigger a payment once certain conditions are met, such as the delivery of a certain quantity of goods. Another benefit of using blockchain technology in supply chain management is the ability to track products and ingredients throughout the entire supply chain. This can help improve traceability and transparency, which is increasingly important for consumers who want to know more about where their food comes from and how it was produced. While there are still challenges to be addressed in implementing blockchain technology in supply chain management, such as interoperability between different systems and the need for standardization, it is a promising solution that has the potential to revolutionize the way we manage and track supply chains.

U.S. Drug Supply Chain Security Act of 2013 mandates that pharmaceutical companies identify and trace prescription drugs to protect consumers from counterfeit and harmful products. Blockchain technology is very specific to data security and traceability. Thus, companies in the pharmaceutical industry can use blockchain to effectively track drug inventory and meet regulatory compliance. All data attached to the inventory transaction will provide important details about the drug and adding the regulatory agency as a trusted partner provides full disclosure of the product's origin and its journey to the consumer. Organizations can benefit from taking quick corrective actions when they discover faulty products. Blockchain traceability provides in-depth data that identifies all parties that are involved in the transactions; therefore, identifying the exact batch where the tainted product may have been procured creates a huge advantage for all stakeholders. Standardization is another important factor to consider. Such as agreeing to centralized data regarding production and inventory allocation decisions onto one repository. While standardization in the supply chain system will ultimately create trust, and transparency, and drive value for organizations, it is not realistic for companies that are involved to accept centralized decisions.

However, the use of blockchain technology to share inventory flows is simple and provides each company with the autonomy to execute their decision-making and the level of control to share the data for visibility purposes. Overall, the use of blockchain technology in the pharmaceutical industry can significantly improve supply chain security and compliance with regulations. By tracking drugs at every stage of the supply chain, blockchain provides transparency and traceability that helps identify potential issues and allows for quick corrective actions. The use of a trusted partner, such as a regulatory agency, adds an extra layer of assurance for consumers and stakeholders. Standardization is important in any supply chain system, but it may not always be feasible for companies to accept centralized decisions. Blockchain technology allows for sharing of inventory flows while maintaining autonomy and control over decision-making. This provides companies with the flexibility to operate their systems while still benefiting from the transparency and traceability provided by blockchain technology. The pharmaceutical industry can greatly benefit from the implementation of blockchain technology in its supply chain systems. It provides a secure and efficient way to track drugs and maintain compliance with regulations while allowing for flexibility and control for each organization involved.

Supply chain giant Walmart is a pacesetter in blockchain technology and has implemented several facets of logistics into blockchain-based supply chains. Some benefits that Walmart has realized through blockchain technology are synchronized logistics, data, shipment tracking, and automation of payments. Blockchain's appeal is to enhance supply chain efficiencies by increasing the speed of recording transactions, improving traceability, and allowing access to all participants on blockchain technology to share data that is immutable.

Conclusion

A transaction on a blockchain typically involves two parties who initiate the transaction by sending data to the network. This data is then encrypted by the blockchain using cryptographic algorithms, which ensures that the transaction is secure and cannot be tampered with. Once the transaction data is encrypted, the blockchain verifies the transaction data and validates it to ensure that it conforms to the rules and protocols of the blockchain network. If the transaction is valid, a new block is created and the transaction data is stored on that block, along with other validated transactions. As more transactions are recorded, new blocks are added to the blockchain, forming a chain of blocks that contains a complete history of all transactions that have taken place on the network. This chain of blocks is stored and replicated across a network of computers, making it virtually impossible to manipulate or alter the transaction history. The adoption of blockchain technology has been slower in some industries, but many organizations are realizing the true value of this technology in providing secure, transparent, and decentralized systems for a wide range of applications. Some reported success cases include supply chain management, financial services, healthcare, and more.

The COVID-19 pandemic has highlighted the importance of supply chain management and exposed its weaknesses. The pandemic has caused disruptions in global trade and supply chains, leading to shortages of essential goods such as medical supplies, food, and other consumer goods. The pandemic has affected the entire supply chain, from raw materials to production to shipping and delivery. Lockdowns and travel restrictions have led to closures of factories, ports, and warehouses, disrupting the flow of goods. Furthermore, panic buying and stockpiling by consumers have put additional pressure on supply chains, leading to shortages of certain products. The COVID-19 pandemic has highlighted the need for more resilient and flexible supply chains. Companies need to have contingency plans in place to deal with unexpected disruptions in the supply chain, such as natural disasters, geopolitical tensions, and pandemics. This may involve diversifying suppliers, increasing inventory levels, or implementing digital technologies to improve supply chain visibility and agility. Thus, the COVID-19 pandemic has shown that a weak supply chain can wreak havoc and cause significant disruptions to global trade and commerce. It has also highlighted the need for companies and governments to invest in building more resilient and sustainable supply chains to ensure that essential goods are available to consumers, even during times of crisis.

In supply chain management, learning is an essential component for ensuring that the supply chain is operating effectively and efficiently. The process of supply chain learning involves a set of procedures that aim to promote learning and knowledge sharing throughout the supply chain.

The procedures for promoting supply chain learning include:

1. **Training and development:** This involves providing training and development opportunities for individuals and teams within the supply chain to improve their knowledge and skills.
2. **Information sharing:** This involves the sharing of information and knowledge between different parties within the supply chain, including suppliers, manufacturers, distributors, and retailers.
3. **Benchmarking:** This involves measuring and comparing performance metrics and best practices within the supply chain to identify areas for improvement.

4. **Continuous improvement:** This involves continuously improving the supply chain processes and operations through a structured process of monitoring, analyzing, and implementing changes.
5. **Governance:** This involves the management and governance of the supply chain processes to ensure compliance with regulations and industry standards.

Therefore, the supply chain learning process aims to continuously sustain knowledge and improve the performance of the supply chain through the management of processes and measuring benchmarks. By promoting learning and knowledge sharing throughout the supply chain, organizations can improve their operational efficiency, reduce costs, and enhance customer satisfaction.

The framework in this chapter substantiates a built-in approval workflow for both the employer, and the contractor provides a single source of validation for the contractor's timesheet that is supported by blockchain technology. How would this work?

1. Purchase orders, time records, and invoices are put on the platform through the utilization of existing systems records.
2. Contractors are matched through the purchase orders.
3. Time and expenses are tracked against the purchase order (this mechanism has a predefined function for instance if the PO states 40 hours, then the time tracked by the system matches against the PO to fulfill this criterion).
4. Once consensus is reached, transactions are recorded on the blockchain ledger.
5. This is followed by invoice generation, reconciliation, and payment.

Blockchain technology has the potential to revolutionize supply chain management by providing a secure, transparent, and decentralized platform for tracking and verifying transactions. By using blockchain, supply chain participants can create an immutable record of every transaction, from sourcing raw materials to the final delivery of goods to customers. This allows for greater visibility and accountability throughout the supply chain, reducing the risk of fraud, counterfeiting, and other forms of malfeasance.

Blockchain technology can also improve the efficiency and speed of supply chain processes by automating many of the manual tasks that are currently performed by humans. For example, smart contracts can be used to automate the verification and approval of transactions, reducing the need for intermediaries and speeding up the flow of goods and services.

It is important to note that implementing blockchain technology in supply chain management is not without its challenges. One of the main challenges is interoperability—ensuring that different blockchain platforms can communicate with each other seamlessly. Additionally, there may be issues with scalability, privacy, and regulatory compliance that need to be addressed. The benefits of blockchain technology in supply chain management are significant, and many organizations are already exploring its potential to improve their operations and reduce costs.

Chapter Summary

Blockchain is a distributed ledger technology that allows for secure, transparent, and decentralized record-keeping, making it well-suited for use in a variety of applications. One area where blockchain technology is being used is in the supply chain management of goods and services. With blockchain, companies can create a transparent and secure record of the movement of goods from one party to another, making it easier to track and verify transactions. This can help to reduce fraud and errors in the supply chain and improve efficiency. In terms of urban governance, blockchain technology can be used to create more transparent and accountable systems for voting, land registry, and other municipal functions. By using blockchain, governments can create secure and tamper-proof records of transactions and decisions, which can improve trust and accountability.

In the industrial economy, blockchain technology can be used to create more efficient and secure systems for managing supply chains, tracking products, and managing contracts. This can help to reduce costs and increase transparency, which can be particularly important in industries like manufacturing, where supply chain complexity can be a major challenge. Blockchain technology can also be used to promote ecological livability. For example, blockchain can be used to create secure and transparent systems for tracking carbon credits or managing renewable energy grids. This can help to incentivize sustainable practices and reduce the impact of climate change. Blockchain technology has a wide range of potential applications in many different areas, and its importance is likely to continue to grow as more organizations explore its possibilities.

In summary, blockchain technology can add value in the realm of supply chain management in three major areas:

1. **Shifting from legacy technology, manual processes, and consolidation:** Blockchain technology supports the potential to replace outdated legacy and manual systems by streamlining business operations.
2. **Reinforcing traceability:** Traceability is a crucial element of supply chains, especially in industries where consumers demand information about the origin of the products.
3. **Reduction in costs:** There is a vast volume of data generated through supply chains which leads to challenges for organizations to safely store and manage the data. While traditional systems struggle to handle the large scale of the data generated, blockchain's distributed ledger function addresses this issue.

In addition, blockchain-based databases offer a range of benefits for agricultural and other supply chains. Here are some of the ways in which blockchain can support the traceability and management of agricultural products:

1. **Information storage:** Blockchain-based databases can provide a secure and decentralized way of storing information related to agricultural products. This information can include data about the product's origin, production practices, quality, and more.

2. **Traceability:** By using blockchain technology, it is possible to create a tamper-proof record of each step in the supply chain. This means that it is possible to track the movement of products from farm to table, providing greater transparency and accountability.
3. **Smart contracts:** Smart contracts are self-executing contracts with the terms of the agreement directly written in the code. This means that once the conditions are met, the contract is automatically executed. Smart contracts can be used in agricultural supply chains to automate payments and ensure that all parties in the supply chain are meeting their obligations.
4. **Dynamic data storage:** Blockchain technology allows for the storage of dynamic data, which means that the database can be updated in real-time as new information becomes available. This can help to reduce the time and cost associated with manual data entry and provide more accurate and up-to-date information for supply chain management.

Therefore, blockchain-based databases offer a range of benefits for agricultural supply chains, including increased transparency, traceability, and efficiency. By leveraging this technology, it is possible to create a more sustainable and resilient food system.

Take Away Lesson from the Chapter

1. Blockchain technology is currently at the heart of goods and services, urban governance, industrial economy, and ecological livability.
2. Some of the most appealing characteristics of blockchain are speed, transparency, and accessibility. These features of a blockchain are very attractive as solutions to legacy systems currently managing masses of data that are fragmented.
3. Smart contracts use symmetric key encryptions. Two types of transactions exist in a smart contract: incoming transactions and stored transactions.
4. The supply chain learning process includes a set of procedures to promote supply chain learning, operating, and governing to continuously sustain knowledge through the management of processes and measuring benchmarks.
5. The blockchain platform offers a built in consensus layer where disputes are settled based upon agreed rules that ensure accuracy, fairness, and time efficiency for all participants.
6. Important features of the blockchain that create value in information traceability are achieved through timestamps, consensus algorithms, data encryption, and smart contracts which are all components of a blockchain.
7. Operational efficiency is another advantage of the blockchain technology application in supply chain management, this can be achieved by the incorporation of smart contracts that build automatically executable interactions between blockchain users.

Chapter Exercises

1. Think about a problem that organizations are facing today. What are the effects, or symptoms, of the problem? As a series of "what caused this" questions, what do you think is the root cause of the problem and do you think blockchain can resolve the issue, if so, how?

2. As you were reading Chapter 5, what did you learn from the framework and what is the most useful information received from the chapter framework and why?

3. What are the main benefits of blockchain based technology in supply chain management? Why do you think that blockchain technology can be impactful to business processes?

PART II

Implementation, Advantages and Disadvantages of Blockchain

Blockchain and Structural Implementation

Big data has been the lifeblood of innovations and has led the information technology industry to innovate new technologies to use big data in a meaningful way. Data security and integrity have been at the forefront of technological innovations. The protection of data privacy and privacy leakage problems creates great security risks for organizations. In addition, when data is forged from data files stored in a database, due to the storage of large amounts of data, it is difficult to detect when data is compromised or tampered with. Blockchain technology creates an immutable and decentralized environment mitigating tampering with data. Basic blockchain theory interlocks smart contracts, consensus mechanisms, block structures, and group signature algorithms at the foundation of the technology. Peer-to-peer network—in a peer-to-peer network, there is no central authority therefore there are no central system standards. Transactions and blocks in a blockchain are transmitted between participating nodes. Once the blockchain receives that data, it verifies the transaction's validity. If validated, it is accepted. The utilization of blockchain radically improves data integrity, and openness, and promotes the execution of trustworthy, transparent, safe, and democratized services and applications. Access rights determine what type of blockchain to use. This chapter prudently disseminates the importance of big data, and current technology flaws, and highlights the usefulness of blockchain technology as the next-generation solution for big data storage, access, and analysis to bring value to organizations and consumers.

Introduction

The plethora of information available on the internet poses several critical issues. The lack of control of one's data causes privacy issues. Thus, the collection and protection of data are at the heart of organizations' minds. To understand how data moves in a common IoT environment we need to understand the flow of data. IoTs can be divided into three types: sensor node, gateway node, and the user-side control terminal. Sensor node: collects data and executes instructions. Gateway node: uploads data collected by the sensor node.

User-side control terminal receives data from the gateway node and allows users to control the IoT devices. The flow of data in an IoT environment typically starts with the sensor node, which collects data from various sources such as sensors, cameras, or other connected devices. The collected data is then processed and analyzed at the sensor node, which may involve filtering, aggregation, or other data processing techniques. Once the data has been processed, it is transmitted to the gateway node, which serves as a bridge between the sensor node and the user-side control terminal. The gateway node may perform additional processing or filtering of the data before sending it to the control terminal.

Finally, the user-side control terminal receives the data from the gateway node, where it can be viewed, analyzed, and used to control IoT devices. The control terminal may also send commands or instructions back to the sensor node or gateway node, which can be used to modify the behavior of the IoT devices. In terms of data privacy, organizations need to implement appropriate security measures to protect the data as it moves through the IoT environment. This may involve encrypting the data as it is transmitted, ensuring that the devices are properly authenticated and authorized, and implementing other security measures such as firewalls, intrusion detection systems, and access controls. Additionally, organizations should ensure that users are aware of the data being collected and how it will be used and provide clear options for opting out of data collection if desired.

The user-side control terminal is the entire system and performs services of operation on its entities. The internet was created and developed for military use. Advancements on the internet have gradually spread from singular military use to civilian use and now serve as the backbone of businesses (Wang, 2022). The protection of data privacy and problems of privacy leakage create great security risks for organizations. When data is forged from data files stored in a database, due to the storage of large amounts of data, it is difficult to detect when data is compromised or tampered with. Blockchain technology creates an immutable and decentralized environment mitigating tampering with data.

Basic blockchain theory interlocks smart contracts, consensus mechanisms, block structures, and group signature algorithms at the foundation of the technology. In a peer-to-peer network, there is no central authority therefore, there are no central system standards. Transactions and blocks in a blockchain are transmitted between participating nodes. Once the blockchain receives that data, it verifies the transaction's validity. If validated, it is accepted.

Consensus mechanism—there are no leaders or boards in a blockchain. Decision-making on a blockchain is achieved through a consensus mechanism, which is a way of reaching a common agreement through consensus that ensures trust in the blockchain is achieved for the benefit of the entire network. Cryptography-enabled secure communication that guarantees the confidential and verifiable integrity of the message.

Organizations are facing serious problems relating to data security. These serious issues include data fragmentation, effective data sharing, and data privacy. Data fragmentation means it is difficult to integrate stored data. Data sharing among all users is inadequate. Data protection and privacy and authority over data access are also a concern. Disasters cause serious business disruptions. The world has recently experienced disruptions caused by the COVID-19 pandemic. There is a crucial need for establishing a system that shares and disperses information effectively. Security vulnerabilities, data privacy concerns, and lack of transparency are some major factors that affect information-sharing applications (Hu et al., 2022).

Currently, organizations rely upon cloud-based storage systems to store consumer data. Data stored in cloud-based servers encounter data integrity, authentication, and privacy violations. Since the data is managed centrally, data can easily be stolen, manipulated, or altogether deleted. In addition, cloud-based storages ignore consumer authority and privacy.

Consumer data on the web is fragmented. Consumers visit billions of sites and use a multitude of different applications where they share data. Since consumers share sensitive data with many organizations, systems that facilitate the sharing of this sensitive data must give priority to consumers' right to privacy. Consumers should be empowered to decide who has access to their data.

Current cloud-based storage systems can pose risks to data integrity, authentication, and privacy. Centralized storage systems like these can be attractive targets for hackers or malicious actors seeking to steal or manipulate data. Additionally, when data is stored in the cloud, it may be subject to data breaches or other security vulnerabilities. To address these concerns, organizations can implement various security measures to protect their cloud-based data, such as data encryption, multi-factor authentication, and access controls. These measures can help ensure that only authorized individuals can access and modify data, and that data is protected from unauthorized access or manipulation.

Furthermore, organizations should prioritize consumer privacy when designing and implementing data-sharing systems. Consumers should have control over their data and should be able to decide who has access to it. This can be accomplished using privacy settings and consent mechanisms that allow consumers to choose which data they wish to share and with whom. In addition, organizations should be transparent about their data collection and sharing practices and should provide clear and easy-to-understand privacy policies. This can help consumers make informed decisions about what data they share and with whom and can help build trust between consumers and organizations.

Blockchain is a feasible solution for consumer data security. A blockchain is a tamper-resistant, affix-only chain of records that are stored in nodes in a network. Blockchain offers public or private use to maintain consistent records through numerous machines all generated through a trusted third party. The use of smart contracts embedded in a blockchain can indeed be an effective methodology for sharing confidential data. Smart contracts are self-executing contracts with the terms of the agreement between buyer and seller being directly written into lines of code. The use of smart contracts in a blockchain network provides several advantages, such as automated verification, execution, and enforcement of the terms of the contract, reducing the need for intermediaries, and increasing transparency and immutability. For instance, in the financial services industry, the use of smart contracts in blockchain technology has led to the emergence of decentralized finance (DeFi) platforms. DeFi platforms aim to replicate various activities of the traditional financial system using blockchain and smart contracts, such as lending, borrowing, and trading of digital assets, without the need for intermediaries.

Industry experts Edmonds (2023) assert that there is a huge explosion of digital technology such as machine learning, cloud computing, blockchain and distributed ledger technology, Web 3.0, and smart contracts that is disrupting industries, especially the financial services sector, and have given genesis to DeFi that is enabling disruption in financial space. DeFi is among the latest trends to be adopted to replicate various activities of the traditional financial system. DeFi has gained significant attention from investors and users alike, as it provides a more open, transparent, and permission less financial system. However, it is important to note that DeFi is still a relatively new and evolving technology, and there are still challenges such as regulatory uncertainty, security concerns, and the potential for market volatility. Therefore, the use of smart contracts embedded in a blockchain can be an effective methodology for sharing confidential data, and it has also led to the emergence of new disruptive technologies such as DeFi in the financial services industry. While there are still challenges to overcome, the potential benefits of blockchain and smart contract technology are significant and are likely to drive further innovation and disruption across various industries in the future.

Blockchain is a distributed ledger technology that features decentralization, transparency, democracy, security, and immutability. The utilization of blockchain radically improves data integrity, and openness, and promotes the execution of trustworthy, transparent, safe, and democratized services and applications. Access rights determine what type of blockchain to use. The four main types of blockchains are public, private, consortium, and hybrid.

Blockchain is a digital database that is transparent, distributed, immutable, decentralized, and operates through parity by allowing access to the public (Makani et al., 2022). All participants on a blockchain network are given access and authority to authenticate the transactions that exist on that blockchain. Blockchain files data into blocks, these blocks are chained together and communicate with each other therefore the name blockchain. Every block on a blockchain consists of several transactions and each transaction is represented in Hash form. A Hash is a code that is a unique identifier embedded during the development of the block, a sort of DNA. If there are changes that are made to the block, the Hash is altered, and this creates a warning mechanism for others on the blockchain.

Blockchain has several other unique properties making it an attractive technology for various applications. One of the most notable features is the fact that blockchain is decentralized, which means that there is no need for a centralized authority or intermediary to manage transactions. This makes blockchain particularly useful for applications where trust and security are critical, such as financial transactions and supply chain management. Another important feature of blockchain is its transparency. Since all transactions are recorded on the blockchain and can be viewed by anyone with access, there is a high degree of transparency and accountability in the system. This can help to prevent fraud and other types of misconduct, as well as provide a greater level of transparency and visibility into how transactions are processed and managed. Blockchain is also highly secure and resilient. Because the data on the blockchain is stored across a distributed network of nodes, it is much more difficult for hackers or malicious actors to compromise the system. Additionally, since each block on the chain is linked to the previous block, it would be extremely difficult for an attacker to alter or manipulate the data on the blockchain without being detected. Ultimately, blockchain is a powerful technology that has the potential to transform many industries and applications. Its unique combination of decentralization, transparency, and security makes it a promising solution for a wide range of use cases.

Structure

The three main components of a blockchain are node, block, and transaction. A node is a device or user that has access to a complete reproduction of the blockchain ledger. A block is a mechanism to store the data structure. A transaction is the minutest component of the blockchain system. Each transaction is organized in the form of a block. Each block is then distributed to every node in the blockchain network. When a new transaction is added to the network, each node validates and processes the new transaction independently. Consequently, each block consists of many transactions that are shared in the network. The composition of the block consists of data that is in the format of a block header, the hash value of the current block, and the hash value of the previous block. Blockchain uses cryptographic algorithms to generate hash values. The use of cryptographic algorithms ensures easy identification of blocks on the blockchain network. As new blocks are created, a hash value is automatically generated and the nomenclature for future blocks follows the same pattern of association. The hash of a block is a crucial element of the blockchain because it creates a chain of blocks. Any alteration of the block creates warnings for the users that must be resolved.

Centralized or Decentralized Blockchain

Blockchain distributed ledger function creates access to all on the blockchain. However, access can be controlled through the application of a centralized or decentralized blockchain structure. For instance, a private blockchain structure is inherently more centralized as it is controlled by a group for increased privacy. Decentralized blockchain are open-ended and open to the public. All records on a decentralized blockchain are open to all participants to take part in the agreement process. One of the disadvantages associated with decentralized blockchain mechanisms is time efficiency. Since the decentralized blockchain is open to all users, it requires massive computation power to maintain the architecture.

Peer-to-Peer Network

Blockchain is a consensus-driven peer-to-peer network where all peers share equal access and participate in transactions through consensus algorithms. This type of network offers a higher level of security because there are no mutual points of failure as compared to a centralized system. The purpose of the peer-to-peer network is to achieve a decentralized, democratized system.

Consensus Algorithm

The consensus algorithm that drives blockchain technology is a core part of the technology. Each participant on the blockchain must follow the rules of the algorithm. Since there is no distinct entity that verifies the transactions, participants' acceptance of the consensus algorithm enables each transaction to be verified and secured on the blockchain. The consensus algorithm ensures that each participant is operating in the latest most synchronized version of the local blockchain structure. Proof-of-Work (PoW) and Proof-of-Stake are common consensus algorithms utilized on blockchain networks. Originally used for the creation of Bitcoin, the Proof-of-Work algorithm validates transactions and then generates new blocks on the blockchain. It takes approximately 10 minutes to create a new block. The nodes that validate the transactions are rewarded by hash algorithm-based mathematical calculations that are solved. The amount of computation that is required to validate is consequently positioned to be completed through a rigorous process to avoid data forgery.

A consensus algorithm is a process or set of rules that allows participants in a distributed network to agree on the current state of the network's shared data. In simpler terms, it is a method used by blockchain networks to reach an agreement on the validity of transactions and the state of the network without the need for a central authority. In a blockchain network, transactions are verified and added to the blockchain by nodes, which are distributed computers connected to the network. Each node maintains a copy of the blockchain, which contains a record of all transactions in the network. Consensus algorithms ensure that all nodes on the network agree on the state of the blockchain, i.e., the order and validity of transactions.

Different blockchain networks use different consensus algorithms, each with its own set of advantages and disadvantages. Some of the most popular consensus algorithms include Proof of Work (PoW), Proof of Stake (PoS), Delegated Proof of Stake (DPoS), and Byzantine Fault Tolerance (BFT). Proof of Work (PoW) is the consensus algorithm used by the first and most well-known blockchain network, Bitcoin. In a PoW-based network, nodes compete to solve complex mathematical problems (What is Proof of Work (PoW) in Blockchain?, 2023), and the first node to solve the problem adds the next block to the blockchain. This process is resource-intensive, requiring significant computing power, and can lead to high energy consumption and transaction fees.

Advantages of Proof of Work (PoW):

1. **Security:** PoW is known for its security as it requires a significant amount of computational power to perform a 51% attack on the network. This means that it is difficult for any single entity to control the network, ensuring the integrity of the blockchain.
2. **Decentralization:** PoW-based networks are highly decentralized, with no single entity controlling the network. This makes it difficult for any government or authority to censor transactions or control the network.
3. **Fairness:** PoW ensures that transactions are verified and added to the blockchain in a fair and transparent manner. All nodes in the network have an equal chance of adding new blocks to the blockchain, making it difficult for any single entity to manipulate the network.

Disadvantages of Proof of Work (PoW):

1. **High Energy Consumption:** PoW requires significant computational power, leading to high energy consumption and associated costs. This has led to concerns about the environmental impact of PoW-based networks.
2. **Slow Transaction Processing:** The process of solving complex mathematical problems to verify transactions can be slow and lead to higher transaction fees, making it difficult for PoW-based networks to handle high transaction volumes.
3. **Centralization of Mining Pools:** Mining pools have become increasingly popular in PoW-based networks, allowing miners to pool their resources and share rewards. This has led to a concentration of mining power in a few mining pools, which can potentially lead to a 51% attack.

Proof of Stake (PoS) is a newer consensus algorithm that uses a different approach to reach consensus. In a PoS-based network, nodes do not compete to solve complex mathematical problems. Instead, they are chosen to validate transactions based on the amount of cryptocurrency they hold in their wallet (What Does Proof-of-Stake (PoS) Mean in Crypto?, 2023). This approach is less resource intensive than PoW, leading to lower energy consumption and transaction fees.

Delegated Proof of Stake (DPoS) is a variant of PoS, where token holders vote for a select group of nodes to validate transactions on their behalf. This approach is faster than PoW and PoS, but it also requires a higher degree of trust in the elected nodes.

Advantages of Proof of Stake (PoS)

1. **Lower Energy Consumption:** PoS consumes significantly less energy than PoW as it does not require nodes to solve complex mathematical problems. This means that PoS-based networks are more environmentally friendly and less costly to operate.
2. **Faster Transaction Processing:** PoS allows for faster transaction processing and can handle higher transaction volumes than PoW-based networks. This makes it more suitable for use in applications that require fast transaction processing, such as payment systems.
3. **Decentralization:** PoS-based networks are highly decentralized, with no single entity controlling the network. This makes it difficult for any government or authority to censor transactions or control the network.

Disadvantages of Proof of Stake (PoS)

1. **Centralization of Wealth:** PoS rewards nodes with the most cryptocurrency, leading to a concentration of wealth in a few nodes. This can lead to a centralization of power, which can potentially lead to a 51% attack.
2. **Lack of Fairness:** PoS rewards nodes with the most cryptocurrency, leading to a lack of fairness. This means that nodes with more cryptocurrencies have a greater chance of being chosen to validate transactions than those with fewer.
3. **Potential for Staking Attacks:** PoS networks are vulnerable to staking attacks, where an attacker acquires a significant amount of cryptocurrency and uses it to manipulate the network. This can lead to a loss of trust in the network and a decline in its value.

Byzantine Fault Tolerance (BFT) is a consensus algorithm (Daly, 2021) that aims to achieve consensus in a distributed network even when some nodes are malicious or faulty. BFT-based networks require nodes to communicate with each other and agree on the state of the network, even if some nodes are intentionally trying to disrupt the network. Thus, consensus algorithms are a critical component of blockchain technology, allowing nodes in a distributed network to agree on the validity of transactions and the state of the network. Different consensus algorithms have their advantages and disadvantages, and the choice of algorithm depends on the specific needs of the network.

Advantages of Byzantine Fault Tolerance (BFT):

1. **High Fault Tolerance:** BFT is designed to tolerate the failure of some nodes or actors in the network. This means that even if a certain number of nodes or actors fail, the network can continue to operate without interruption.
2. **Faster Transaction Processing:** BFT-based networks can process transactions quickly and efficiently, making them suitable for applications that require fast transaction processing, such as payment systems.
3. **Decentralization:** BFT-based networks are highly decentralized, with no single entity controlling the network. This makes it difficult for any government or authority to censor transactions or control the network.

Disadvantages of Byzantine Fault Tolerance (BFT):

- **Higher Complexity:** BFT is a more complex consensus algorithm than PoW or PoS, which can make it more difficult to implement and maintain.
- **Higher Cost:** Implementing a BFT-based network can be more expensive than other consensus algorithms due to the need for additional resources to ensure fault tolerance.
- **Potential for Attacks:** While BFT is designed to tolerate the failure of some nodes or actors in the network, it is vulnerable to attacks by a group of malicious nodes or actors. These attacks can compromise the integrity of the network and lead to a loss of trust in the network.

Proof-of-Stake Algorithm

Since there is a substantial time commitment for validating transactions and creating blocks, the Proof-of-Stake algorithm provides a forum where some nodes in the blockchain network create only new transactions and not all. Therefore, these specific nodes that validate transactions in a Proof-of-Stake algorithm are known as validators. The validators increase their stake while creating a new block. Therefore, the validators with the highest value prevail while carrying out the new transactions. During the creation of a new block, its coinage is reset to zero, giving other nodes priority over the current node to generate the new block. By resetting the priority of nodes, the dominance of the nodes reduces the large stake.

Digital Identity

Digital identity is the data that identifies an individual or an entity. Users use a variety of databases and share their information with these databases. Digital identity is a reservoir of information that is shared in different databases and is becoming a focus for individuals and organizations to take ownership of their data and data-sharing rights in a controlled and realistic fashion.

Finance–Problem–Trust and Transparency

Blockchain is a viable solution for the financial sector. The fact that blockchain offers one ledger that can be shared across multiple computers through private or public access provides a great deal of trust and transparency to the users. The achievement of transparency with all stakeholders is extremely valuable. Utilizing blockchain in finance allows transparency to all users of the blockchain. Frontline workers, supervisors, managers, leaders, board of directors, clients, and regulators. Blockchain provides all the stakeholders with the most updated information on transactions that concerns them. The single ledger format makes it easy for workers to complete financial reports and hold all financial transactions and related dates of the organization. The use of smart contracts is another sustainable solution since smart contracts are based on facilitating, verifying, and enforcing contracts. For instance, in real estate transactions, a smart contract is executed with a base set of rules for the jurisdiction that involves the buyer, the seller, the financial institution(s), and the government agency. All built-in rules in the contract are enforced automatically. The smart contract system operates on a set of codes, once the parameters defined in the contract are met, a code is generated and provided to the financial institution as a form of payment to be unlocked through the generated code. The usage of smart contract technology is a feasible application in banking by creating automated, verified transactions between parties and eliminating legacy time-consuming approval systems that are manual.

Privacy Protection

Li et al. (2018) used an online platform for data sharing and discovered that RZKPB. RZKPB is a privacy-preserving blockchain-based fair transaction method for sharing economy that is a more efficient format of the blockchain-based privacy-preserving method. The problem between privacy preservation and efficiency can be resolved through the RZKPB application. The integrity of financial transactions on a blockchain is maintained by sharing specific records regarding the payer, payee, date, amount, and services to be executed. A centralized solution of blockchain-based technology allows peer-to-peer transparency and immutability to financial transactions, eliminating the need for a third party and creating efficiencies. Smart contracts are an essential part of the data creation, management, and storage of the future. Blockchain technology can house loans, credit information, financial exchanges, and exchanges with the government to properly record these real transactions when needed. Blockchain-based databases facilitate the production of computerized record exchanges through the application of cryptography.

Financial institutions transact with money. Consumers have always desired more transparency. The adoption of blockchain technology accomplishes transparency in the financial sector. In retail banking, there are many human errors that occur because of poorly trained and managed employees. Blockchain offers a concrete solution to remove human error in banking. Through the use of blockchain technology, all financial transactions are available on one ledger for easier adjustments, balancing, and compilation. In the financial sector, a critical operating factor in the trading of securities is based upon Know your Customer (KYC). Regulators have instigated several initiatives to manage KYC among the various industry players due to severe shortfalls that have created many Anti Money Laundering AML issues in the past. The premise of opening an investment account is to know critical data points about the client. Firms like JP Morgan Chase have spent 1.6 billion euros to contain KYC for the firm (Underwood, 2016). Blockchain technology provides a feasible resolution for KYC, where clients upload data to verify their identity, thereby creating substantial cost savings for the industry. Money flows that cross border transactions create more inefficiencies. The Society of Worldwide Interbank Financial Telecommunication

(SWIFT) is a banking service that provides the execution of international financial transactions. In most cases, SWIFT transfers can take 2-5 days, and this occurs due to the existence of a middleman generally known as a clearing agent. The inclusion of blockchain technology removes the middleman and creates time and cost efficiencies.

Application

The implementation of blockchain technology in different sectors of the market can provide substantial advances in globalization by enabling virtual transactions to be concluded in a decentralized and secure manner, managing numerous smart contracts, and contributing to cost efficiencies (Rodrigues and Garcia, 2023).

Global industries are undergoing rapid changes to accommodate massive developments in recent years that are enabled by globalization and boosted by policy and incentives. Industries are consequently looking for evolutionary and sustainable technology to create efficiency. Currently, tasks are increasingly challenging in many industries. A need for moderate smart management and control is required to alleviate the growing need for industry transactions by decentralizing and removing complex and multi-agent actors as the intermediary (Guo et al., 2016). Blockchain technology is primarily designed to facilitate distributed transaction by removing central management and can assist in the distribution of local control and management through decentralization. Additionally, blockchains enable shared and distributed data structures that securely store digital transactions without one central point of authority. More distinctively, blockchains permit the automation execution of smart contracts in peer-to-peer networks (Palizban et al., 2014). Therefore, blockchain databases permit multiple users to make changes in the ledger simultaneously, instead of managing the ledger with a single trust source. Incorporating multiple network members who hold a copy of the recording chain and reaching an agreement on the valid state of the ledger through a consensus mechanism increases trust and transparency. The use of cryptography enables each new transaction to be linked to the previous transaction making the architecture resilient and secure. Furthermore, every user has the autonomy to check for themselves if the transaction is valid, this provides transparency, and trust and builds a tamper-proof environment. Many participants in various industries uphold that blockchain has the potential to radically improve cost savings and efficiency in the operation of systems and markets (Andoni et al., 2019; Walport, 2016).

Blockchains are digital databases where data transmission is equal to copying data from one place to another. Like in a cryptocurrency domain, this action would equate to copying digital coins from one user's electronic wallet to another. The current databases for information are centralized. This means that there is one heart to the operation that controls it all and creates vulnerability for data security. Organizations are persistently looking for better data security options;

especially as digital data is growing astonishingly. A single cyber-attack on a centralized database creates a complete system compromise and leaves the users defenseless. One of the strengths of blockchain-based databases is decentralization, which allows multiple copies of the same data that are accessible to multiple users on the database. A single attack on the stored data will not be successful as the system is decentralized and has many real-time copies available to different users. Apart from a decentralized ledger, blockchains use distributed consensus algorithms to fortify data security. Mainly run by consensus algorithms that must be resilient to failures of nodes, message delays, corrupt messages, and unreliable and unresponsive malicious codes.

A new block can be generated by nodes in the network. An important step in creating new blocks of data is accepted by the network members in a process called reaching consensus.

- Current market structure—has one center to the operation for instance a bank. The bank has many participants such as the bank (at the center) that employs traders, commercial bankers, retail bankers, etc. in the operations process; however, all processes and structures are embedded and controlled by the central banking database.
- Blockchain market structure—Blockchain structure would create a trading platform for all participants and feed the data back to all users as it is not dependent upon one central hierarchy.

Case Study: Energy

In 2017, Brooklyn, NY-based LO3 Energy Inc. teamed up with Siemens to pilot an energy microgrid using blockchain technology. In this pilot, residents that owned solar panels would be allowed to sell excess energy to their neighbors in peer-to-peer transactions taking advantage of blockchain technology. Energy is volatile, cannot be entirely confined, and is lost through transmission. It is estimated that 5% of electricity created in the US is lost. Energy microgrids can provide an efficient alternative by simply transacting energy stored in already installed solar panels to other members of the local community (Mengelkamp et al., 2018). The LO3 energy grid is called a TransActive Grid that was first tested back in 2016 and combines a meter and a computer that measures both production and consumption, then shares the information so an action can be initiated. This decentralized application platform is cryptographically secure. The blockchain principle of shared ledger which creates visibility for transactions and files them chronologically provides data that is verified and cannot be altered without changing every transaction. Thereby creating a secure platform for both the trader's and consumer's protection.

The TransActive Grid project is clean, reliable, and promises local energy advantages. Some serious concerns were called out due to the super storm Sandy that caused a series of blackouts across the US in 2012. One of the major concerns was the reliability of the energy grid. Most of the country struggles with brutal winter storms that leave power outages mainly because of long transmission lines that are damaged due to extreme weather conditions. The use of TransActive Grid can provide much-needed electricity that can be directed to hospitals, shelters, and community centers as needed.

Photovoltaic panels (PV) are used to produce electricity from sunlight and are inherently different from solar as solar uses thermal systems to produce heat. In the past, utility companies have controlled the production of extra electricity through PVs. Residents that have PVs could sell excess energy back to the utility company but cannot profit from the sale. Generally, the extra power sold back to the utility company is deducted from the customer's total monthly bill; thereby granting the power to the utility companies to control the process (Sharma, 2020). Microgrids remove the utility company as the intermediary and allow the residents to remain in control of their power production (Tan et al., 2022). The Brooklyn microgrid is an example of a big change in the energy industry; however, working with regulators is key in this heavily regulated industry to use the power of blockchain technology to create responsible sources of localized energy that benefit both the energy producers and the communities it serves. Blockchain-based solutions provide a clear benefit for the markets and consumers. Although clearly an extremely beneficial technology, blockchain technology is a disruptive technology for most industries including energy. There are many challenges to achieving market penetration, including legal and regulatory competition barriers.

Blockchain technology has brought new impetus to the future development of many industries. Big data is ceaselessly expanding and the fragmentation in the collection and dissemination process immensely affects people's lives (Wang and Liu, 2022). The concept of big data is the product of the information age that has produced a surplus of data on individuals. Currently, big data is collected and stored in a wide-ranging format and then distributed to computing architectural databases to conduct data processing and integration analysis. Big data is captured, managed, processed, and organized through mainstream software to help organizations make strategic business decisions. The vast accumulation of big data has created value for organizations. A large amount of big data exists in semi-structured and irregular formats. Today big data is being used by various users to accurately position themselves by understanding the preferences of target customers to gain high-quality marketing efforts. Deep integration with cloud computing and big data has contributed to breakthroughs in the scientific and data community. Although big data has expanded the scope of analysis for organizations, the securitization and compilation of big data are much desired to effectively distribute and analyze consumer preferences, and there has been an active search to consolidate big data effectively.

Blockchain technology features a distributed infrastructure and unique computing methodology that creates a secure and verified database structure. Blockchain technology creates authenticity and reliability of information and reduces interference and technical loopholes. The safeguards built into blockchain technology create warning functions for any abnormality and prevent the user from being cheated and property loss. The technical advantages of blockchain technology achieve point-to-point communication remover intermediaries and provide secure and smooth transactions as technical advantages for the implementation of blockchain technology in wide-ranging industry capacity. Blockchain-based databases achieve data security through a bridge between communication terminals and data. Blockchain is inherently transparent in transmission and data storage. In the era of continuous globalization and rapid change in industries, blockchain can uniquely offer strengthened features for government and regulatory supervision. Blockchain technology can greatly promote the dissemination of information especially big data that can strongly support business strategy growth. The main characteristics and values of blockchain technology are transparency, openness, traceability, open sharing, and innovative modules that support auxiliary supervision. The rapid evolution of the information age is surrounded by data that is extremely useful to organizations. Industries are seeking viable use of big data in their industries and as new models change and develop, there are challenges due to data fragmentation and securities and opportunities that can be successfully met through blockchain implementation. Blockchain combines the advantages of big data management to create efficiencies in data analysis and provide valuable insights into various industries. Blockchain improves the ability of various industries to manage and resist risks.

Conclusion

IoT dataflow user-side control terminal receives data from the gateway node and enables the user to control and interact with the IoT devices. The flow of data in an IoT environment typically starts with the collection of data by the sensor node. This data can be anything from temperature readings to motion detection. The sensor node then processes and stores the data locally. Once the sensor node has collected and processed the data, it sends it to the gateway node for further processing and analysis.

The gateway node acts as a bridge between the sensor nodes and the user-side control terminal. It receives data from multiple sensor nodes and aggregates it before sending it to the user-side control terminal. The gateway node is responsible for ensuring that the data is accurate and consistent before it is sent to the user-side control terminal. The user-side control terminal receives the data from the gateway node and presents it to the user in a readable format. The user can then interact with the data and control the IoT devices connected to the network. The user-side control terminal can also send commands back to the gateway node to control the IoT devices. Thus, the flow of data in an IoT environment is complex and requires careful consideration to ensure that data is collected, processed, and transmitted securely and efficiently. Organizations must be aware of the potential privacy issues surrounding IoT data and take appropriate measures to protect user data.

Blockchain technology is built upon the foundation of several key components, including smart contracts, consensus mechanisms, block structures, and group signature algorithms. These components work together to create a decentralized system that allows for secure, transparent, and immutable transactions and data storage. In a peer-to-peer network, there is no central authority or single point of control. Instead, transactions and blocks are transmitted between participating nodes in the network. Each node in the network verifies the validity of transactions and blocks, ensuring that only valid data is added to the blockchain.

- Consensus mechanisms are used to ensure that all nodes in the network agree on the state of the blockchain. These mechanisms can vary depending on the specific blockchain implementation, but they typically involve a process in which nodes work together to validate transactions and add them to the blockchain.
- Smart contracts are self-executing programs that are stored on the blockchain. They allow for automated, trustless transactions between parties, without the need for intermediaries or middlemen. Block structures are used to organize transactions and data on the blockchain. Each block contains a set of transactions, as well as a reference to the previous block in the chain. This creates a tamper-proof record of all transactions that have occurred on the blockchain.
- Group signature algorithms are used to provide privacy and anonymity on the blockchain. These algorithms allow for the creation of anonymous transactions, while still maintaining the integrity and security of the blockchain.

Moreover, blockchain technology is also immutable, meaning that once a transaction is recorded on the blockchain, it cannot be altered or deleted. This feature ensures that the data on the blockchain is secure and cannot be tampered with, providing a high level of data integrity. Another feature of blockchain is its potential for creating smart contracts. Smart contracts are self-executing contracts with the terms of the agreement between buyer and seller directly written into lines of code. This allows for the automation of processes, reducing the need for intermediaries, and increasing efficiency. Blockchain is also highly secure due to its cryptographic protocols that ensure the integrity of the data and protect it from unauthorized access. This makes blockchain a valuable technology for applications that require secure data storage and transfer. Blockchain technology has the potential to increase efficiency and reduce costs in various industries. By removing intermediaries and automating processes, blockchain can reduce transaction costs and increase the speed of transactions. Largely, the unique properties of blockchain make it a promising technology with a wide range of applications, from financial transactions to supply chain management and beyond.

In a blockchain market structure, a consensus is reached through a decentralized process where all nodes in the network must agree on the validity of new blocks before they are added to the chain. This is done through a consensus algorithm such as Proof-of-Work or Proof-of-Stake. Unlike the current market structure where there is a central authority that controls and manages the operations, the blockchain market structure provides a distributed and transparent platform for all participants to trade and exchange data without the need for intermediaries. This allows for a more secure and efficient system as the data is stored on multiple nodes in the network, making it virtually impossible to manipulate or tamper with. Furthermore, the blockchain market structure is not limited to financial transactions and can be applied to a wide range of industries such as supply chain management, healthcare, and real estate. This creates endless possibilities for innovation and disruption as the technology continues to evolve and mature.

Chapter Summary

Blockchain technology has the potential to provide secure and decentralized storage of consumer data. By using a distributed ledger system, a blockchain network can ensure that data is encrypted, tamper-resistant, and only accessible by authorized parties. Moreover, the use of smart contracts in a blockchain network can automate the processing of data transactions while ensuring that the terms of the contract are enforced, and the data is exchanged securely. In addition to providing secure data storage and processing, blockchain technology can also increase transparency and accountability in data management. Since every transaction is recorded on the blockchain, it creates an immutable and transparent audit trail that can be used to verify the integrity of data and track any changes made to it. This can help build trust between consumers and businesses and provide assurance that their data is being handled ethically.

The three main components of a blockchain are node, block, and transaction. Each transaction in a blockchain is verified by multiple nodes in the network to ensure its validity. This is done through a consensus mechanism, which is a way for nodes to agree on the state of the blockchain. There are different types of consensus mechanisms, such as Proof-of-Work (PoW) and Proof-of-Stake (PoS), which use different methods to verify transactions and add blocks to the blockchain. Furthermore, the hash value of a block serves as a unique identifier for that block and all the transactions it contains. Once a block is added to the blockchain, it cannot be altered without affecting the hash values of all subsequent blocks. This makes the blockchain an immutable and secure ledger of transactions. The combination of nodes, blocks, and transactions, along with cryptographic algorithms and consensus mechanisms, enables the blockchain to provide a transparent and secure way of recording and verifying transactions.

In a decentralized blockchain network, no central authority controls the system. Instead, all participants have an equal say in the decision-making process. This makes the system more resistant to attacks, as there is no single point of failure that can be targeted by malicious actors. Decentralized blockchain networks also offer greater transparency, as all transactions are recorded on a public ledger that can be accessed by anyone. On the other hand, a centralized blockchain network is controlled by a central authority, which can make decisions regarding the network's operation. This can result in a more efficient system, as decisions can be made quickly and easily. However, centralized blockchain networks are more vulnerable to attacks, as a single point of failure can bring down the entire network. Both centralized and decentralized blockchain networks have their advantages and disadvantages. Decentralized networks offer greater security and transparency, but can be less efficient, while centralized networks are more efficient but can be more vulnerable to attacks. The choice of which type of network to use depends on the specific needs of the system in question.

In a peer-to-peer network, all nodes or participants have equal rights and responsibilities. This means that no central authority or server is controlling the network. Instead, each node is responsible for storing and sharing data with other nodes. In the case of blockchain, all nodes maintain a copy of the same ledger, and transactions are verified and validated through a consensus algorithm agreed upon by the network. This consensus process ensures that transactions are legitimate and prevents any fraudulent or malicious activity. Because there is no central point of control, peer-to-peer networks are more resistant to attacks and failures. This is because even if some nodes go down or become compromised, the rest of the network can continue to function normally. Additionally, peer-to-peer networks can be more efficient and cost-effective, as they eliminate the need for expensive centralized infrastructure. The goal of a peer-to-peer network is to create a decentralized system where participants have more control and ownership over the network and its data.

A consensus algorithm is a mechanism used in a blockchain network to ensure that all participants agree on the validity of transactions and the state of the ledger. Consensus algorithms enable multiple nodes in a network to come to a mutual agreement on the current state of the blockchain without the need for a central authority. There are several consensus algorithms used in blockchain technology, including Proof-of-Work (PoW), Proof-of-Stake (PoS), Delegated Proof of Stake (DPoS), and Byzantine Fault Tolerance (BFT). Each of these algorithms has its own set of advantages and disadvantages in terms of security, scalability, and energy efficiency. Proof-of-Work is the most widely used consensus algorithm in the blockchain industry, and it requires nodes to perform complex mathematical calculations to validate transactions and add new blocks to the blockchain. Proof-of-Stake, on the other hand, requires nodes to hold a certain amount of cryptocurrency in order to participate in the consensus process. Delegated Proof of Stake is a variation of Proof of Stake that involves the election of a group of trusted nodes to validate transactions. Byzantine Fault Tolerance is a consensus algorithm that allows nodes to reach consensus even in the presence of faulty nodes or malicious attacks. Therefore, the choice of consensus algorithm depends on the specific use case and requirements of the blockchain network.

While blockchain technology has many potential benefits for data security, it is not a silver bullet solution. It is important to consider the limitations and challenges of implementing blockchain-based solutions, such as the high cost of deploying and maintaining a blockchain network, the need for specialized technical expertise, and the potential for regulatory and legal issues. Therefore, it is crucial to carefully evaluate the specific use cases and requirements before deciding to adopt a blockchain-based solution for data security.

Take Away Lessons from the Chapter

1. The lack of control of data causes privacy issues. Thus, the collection and protection of data are at the heart of organizations' forefront decision making.
2. The protection of data privacy and privacy leakage problems creates great security risks for organizations.
3. A consensus mechanism in a blockchain purports that there are no leaders or boards in a blockchain.
4. Cryptography-enabled secure communication that guarantees the confidential and verifiable integrity of the message.
5. The three main components of a blockchain are node, block, and transaction.
6. Blockchain uses cryptographic algorithms to generate hash values. The use of cryptographic algorithms ensures easy identification of blocks on the blockchain network.
7. Blockchain technology is primarily designed to facilitate distributed transactions by removing central management and can assist in the distribution of local control and management through decentralization.

Chapter Exercises

1. What is the difference between decentralized and centralized blockchain technology?
2. What are the issues with big data technology and fragmentation? How do you think blockchain technology fits into a possible solution?
3. What are the main components of blockchain technology and how do they operate? Pick an industry and describe how using blockchain in the industry would be helpful.

Blockchain Opportunities and Vulnerabilities

Public ledger technology has been a part of the history of technological evolution. Blockchain technology is a public ledger of distributed records that facilitates transactions. Early adopters of blockchain technology have experienced breakthroughs in many daily activities and business processes. Blockchain technology can be used for a variety of applications beyond just financial transactions, such as supply chain management, identity verification, or voting systems, among others. Some advantages of blockchain technology include constructing a shared repository of data that is maintained by peers (no third-party intermediary), embracing trust among users through digital signatures and validations ensure every node and user behaves ethically, building potential infrastructure that transforms the next digital fabric of the world by creating a worldwide repository of data that can potentially be accessed by everyone, creating transparency and guaranteeing that everyone on the platform can read the final state of the transactions and have full access to the history of the transaction, automation with infusing smart contracts. Immutability ensures that data cannot be erased or changed, and decentralization creates consensus and removes one central authority. Although there are several advantages of blockchain, there are some notable disadvantages as well including high power consumption that is attributed to the algorithms built in for mining that requires expensive hardware that require a massive amount of computation is required for data mining, since a current copy of the blockchain is available to all users, the data replication requires more space, immutability and transparency eliminate users who do not meet the standards, while a positive for the industry and a negative for the user who had the potential to change business practices but can no longer participate because of some negative data, smart contracts must rely upon external data—this process needs to be either streamlined or the data needs to be injected into the blockchain and, a lack of data governance can cause strong reputational issues and become the weakest part of the process.

Introduction

Blockchain technology features a public ledger that is distributed over a network and records transactions executed among network participants. Each transaction on the blockchain is verified through network nodes and majority consensus before it is added to the blockchain. Recorded transactions on a blockchain cannot be changed or erased. Early adopters of blockchain technology have reported breakthroughs in many daily activities and business processes. Blockchain resembles a long interlocking chain mechanism that connects information. To understand how blockchain technology works, let's use an example of Jack and Jane's need for money transfers. Jack wants to transfer some money to Jane. Jack specifies the dollar amount of the transfer to the network. Jack takes the action of broadcast to the network that he'd like to transfer some money to Jane. The network receives Jack's message about the money transfer to Jane and transfers the requested funds to Jane's wallet. The transaction is validated digitally using a digital identity that is stored in the digital wallet. Digital currency is stored in the form of a digital wallet that cannot be altered by someone else.

Network nodes check the validity of the transaction to ensure that the transaction was authorized by Jack by analyzing Jack's digital signature. Once Jack's digital signature is verified, the nodes then verify if Jack is entitled to spend the money by computing the balance on a local copy of the blockchain. After this transaction between Jack and Jane is verified, the transaction is stored in a block. The new block that is created has a block header, previous block header hash, and transaction hash. Thus, the block contains the list of transactions, and in its header, the hash of the previous block header, and the hash of contained transactions. It's worth noting that this process is possible because of the underlying technology and principles that power blockchains, such as cryptography, decentralization, and consensus mechanisms.

Cryptography allows for secure and private transactions to take place within the blockchain, as each transaction is encrypted and verified using complex algorithms. Decentralization ensures that no single entity has control over the network or the transactions that take place on it, making it more resistant to fraud, censorship, or attacks. In addition, consensus mechanisms like proof of work or proof of stake ensure that the network participants agree on the validity of the transactions before they are added to the blockchain, making it a trustless and transparent system. Furthermore, blockchain technology can be used for a variety of applications beyond just financial transactions, such as supply chain management, identity verification, or voting systems, among others. Its potential uses are still being explored, and its adoption and development are likely to continue to grow in the coming years.

Once the new block is created the mining process begins. The nodes on the blockchain start a competition to solve a mathematical problem finding a random value that, if combined with a block header, produces a given result. The node that won the competition can possibly receive a reward. After the mining process is completed, the new block is then added to the blockchain. The process called proof of work requires nodes to find random values that when combined with the previous block header yield a given result. The proof of work computation is a cumbersome process. One of the reasons why blockchains are tamper resistant is that if a bad actor decided to comprise the blockchain they would have to recompute the proof of work for each block on the blockchain, and it takes hours of computation power to achieve this. Each node updates its local copy of the blockchain and ultimately Jane receives the money that Jack had initiated to transfer to her.

Since its humble beginning in 2008 with the launch of Bitcoin, which was the first of its kind of digital currency traded without relying on intermediaries in an immutable and publicly verifiable manner, Blockchain has evolved tremendously in its potential application (Overstreet et al., 2019).

- **Blockchain 1.0 Era (digital money creation)**—entailed the creation of Bitcoin to the evolution of various cryptocurrencies. More than 600 different cryptocurrencies have developed since Bitcoin.
- **Blockchain 2.0 Era (registering, confirming, and transferring contracts)**—focused on the integration of smart contracts. Smart contracts are algorithm-based codes that are programmed and stored in the blockchain that function on condition (such as if-then logic). Smart contracts automatically execute when the logic is met and do not require the intervention of a third party. In organizational use, smart contracts enable the creation of a decentralized autonomous organization where rules are encoded to make decisions and manage groups of people.
- **Blockchain 3.0 Era (sector application)**—the application of blockchain reaches new heights. No longer tethered to just the creation of cryptocurrency and some finance applications, sectors such as government, healthcare, education and more embrace the technology. A key characteristic of blockchain is immutability which is advantageous to censorship, as people sharing information can do so without someone altering the data.
- **Blockchain 4.0 Era (beyond evolution to adoption)**—blockchain application in real-world scenarios. Healthcare and health sciences can benefit from recorded genomic data (this is currently forbidden in many countries) and make information available to its owners. This critical information can change people's lifestyles by alerting them to a genetic predisposition to a given disease in the foreseeable future. In the education field, data for endowments such as enabling money transfers for students when they pass the course automatically and recording students' outstanding achievements provide transparency in mobility contexts and the job-seeking hiring process. Big data can be utilized to integrate with blockchain technology to create predictive-reactive systems by completing and storing data that can be executed later to make actionable decisions by combining the power of artificial intelligence and smart contracts.

In this book, we focused on four main areas of blockchain application; below is a compilation of how organizations in each of these sectors can benefit from the application of blockchain technology.

1. **Financial Sector:** The capital market structure can be reconfigured and automated for currencies, stocks, private/public equities, crowdfunding, bonds, mutual funds, derivatives, annuities, pensions, insurance policies, trading/spending records, microfinance, charity donations, escrow transactions, bonded contracts, third-party arbitration, multiparty signature contracts, and more. In the financial sector, blockchain has already proven to be very successful. One area where blockchain has yielded significant efficiencies is cross-border payments. Many financial institutes have adopted Ripple to manage international payments. Ripple is a digital currency exchange that provides a scalable, secure, and sustainable central bank digital currency to meet the security standards of central banks.

2. **Healthcare Sector:** Crucial data on patient healthcare management such as patient test results, diagnosis, history, medications, specialty tests and recommendations, and cross-country genome data sharing are all influential factors in better patient care. In the healthcare sector, vital data collection and location of information that transmits critical care information on patients can be achieved through blockchain-based solutions. One of the biggest advantages of relying upon blockchain technology is that it assures that the system will not fail.

3. **Law Enforcement Sector:** Important documents such as driver's licenses, identity cards, passports, voter registration, land/property titles, vehicle registrations, business licenses, business ownership, government records such as marriage, birth, and death certificates, health and safety inspections, gun permits, building permits, and forensic evidence can all be housed on the blockchain. The government has an opportunity to correct information asymmetry in voting by implementing blockchain to gather information regarding the transparency and validity of citizens' votes. When a person votes, there is direct communication with the voter's digital wallet that is validated using a digital signature. By utilizing blockchain functions, voters can cast votes on any digital device guaranteeing anonymity, exclusivity, and immutability.

4. **Supply Chain Management Sector:** Both upstream and downstream data management including all raw material procurement, production and manufacturing, and logistics are incredibly important for a robust supply chain that can be achieved through blockchain technology. The global diamond industry has already incorporated blockchain technology to combat counterfeiting and has experienced positive results. Blockchain allows participants in the diamond industry to trace and locate goods along the supply chain. The traceability function embedded in the blockchain structure creates trust and authenticity without the intervention of a third-party decreasing the possibility of counterfeiting and increasing efficacies in this industry.

Advantages of Blockchain

1. Constructing a shared repository of data that is maintained by peers (no third-party intermediary). All users have access to the data and viewing transactions and storing information on nodes prevents data loss in case of unanticipated events.
2. Embracing trust among users through digital signatures and validations ensures every node and user behaves ethically.
3. Building potential infrastructure that transforms the next digital fabric of the world by creating a worldwide repository of data that can potentially be accessed by everyone.
4. Creating transparency and guaranteeing that everyone on the platform can read the final state of the transactions and have full access to the history of the transaction.
5. Automation with infusing smart contracts.
6. Immutability ensures that data cannot be erased or changed.
7. Decentralization creates consensus and removes one central authority.

Disadvantages of Blockchain

1. High power consumption is attributed to the algorithms built in for mining that requires expensive hardware. A massive amount of computation is required for data mining.
2. As a current copy of the blockchain is available to all users, the data replication requires more space.
3. Immutability and transparency eliminate users who do not meet the standards, while a positive for the industry and a negative for the user who had the potential to change business practices but can no longer participate because of some negative data.
4. Smart contracts must rely upon external data—this process needs to be either streamlined or the data needs to be injected into the blockchain. A lack of data governance can cause strong reputational issues and become the weakest part of the process.

Both advantages and disadvantages of blockchain-based solutions are directly correlated with the organization's needs. Therefore, an organization's own necessities dictate where blockchain is a good solution for its needs. Some questions that organizations must ask while strategizing to shop for a better ethnology such as blockchain are:

1. Do we need a shared database?
2. Do we have multiplayer involved in the data writing process?
3. Do we have potential writers that are untrusted—should we prevent writers from modifying others' previous entries?
4. Is it necessary to remove trusted intermediaries from verifying and authenticating transactions?

5. Is there value in seeing how transactions are linked to each other, or should different users independently write transactions?

In essence, blockchain-based solutions provide an innovative resolution; however, the implementation of blockchain technology is inherently an organizational-based need and must be executed strategically. Blockchain solutions are not a quick or magical fix, they must be carefully considered.

Structure

The global digital payment system has enabled a robust peer-to-peer system regardless of who they are and facilitates receipts and payments anywhere in the world (Mundhe, 2022). Unlike traditional banking systems, the global blockchain-based digital payment system excludes intermediaries and does not rely on banks to authorize the transaction process. Blockchain technology is based on the premise of recording information that is impossible and extremely difficult to modify or hack into. Blockchain books of transactions are duplicated and disturbed throughout the entire blockchain network. Since the genesis of cryptocurrencies, attempts to exploit them illegally through money laundering and breach restrictions have been on the rise, generating the term crypto-jacking. Crypto-jacking is the unauthorized use of one's computer to mine cryptocurrency (Strebko and Ramanos, 2018; Roet It, 2021). Preventing cybercriminal activity has been a main concern for digital currency systems. Blockchain technology helps safeguard against hacking. Blockchain networks store transaction records on many computers in a worldwide peer-to-peer network (Kramer, 2019). A database is a repository for collecting information and storing data on a computer system. The advantages of blockchain technology include a reliable distributed system without the intervention of the government that facilitates optimal security, financial efficiency, better stability, access to real-time data, and cost reduction.

Some of the challenges and limitations currently faced by the finance industry, particularly in relation to the demand side of business transactions. These challenges include income volatility, geographical barriers, information asymmetry, literacy, and lack of trust. The result is that many individuals, particularly those in developing countries or with lower incomes, are unable to access financial services that can support their investment and financial goals. This can contribute to a widening wealth gap and further marginalization of certain populations.

One of the key issues highlighted is the reliance on legacy systems and intermediaries to facilitate financial transactions. These systems can be slow, costly, and prone to errors or information asymmetry, which can erode trust in the system. There is also a need to improve the accessibility and convenience of financial services, particularly for those who are unbanked or who live in areas without easy access to traditional brick-and-mortar banks.

There is a need for innovation and modernization within the finance industry, particularly in terms of leveraging technology to improve efficiency, accessibility, and trust in financial transactions. This could include things like mobile banking, blockchain technology, and other digital solutions that can streamline transactions and reduce costs. Additionally, efforts to increase financial literacy and education could help to bridge some of the gaps in understanding and trust that currently exist between consumers and financial institutions.

Security

Digital assets that are stored on blockchain-based databases are accessed by anyone in the world. Traditionally transactions are processed using intermediaries such as banks, credit cards, and other transfer exchange services (Prasad, 2020). While most individuals reside in developing and stable economies, many people do not have access to banking facilities and are left with no choice but to transact in money and store money in their homes. However, with digital currency, the unbanked can gain access to a much more secure method of saving and transacting for their financial requirements. The security of blockchain technology provides a much better solution than storing cash and cash transactions that create a burden. The safety of digital currency makes it not only safer but easier to use because every transaction that is entered gets recorded and the blockchain network verifies its authenticity. Only upon successful validation of the records, the transaction is confirmed and recorded to the block. If transactions are altered on the blockchain, the hash of the block changes, altering the users of the change (Macedo, 2018).

Transparency

Blockchain technology is an open-source technology where all users on the chain have access to the stored data. For instance, auditors or regulators can review the transaction for security. The lack of one central authority makes the technology transparent for all users. Furthermore, blockchain technology provides faster and easier processing to gain trust as compared to traditional procedures. A digital, paperless system provides more transparency throughout the transaction process for all users. Human error and fraudulent employee behavior can both be avoided by ensuring that all records of transactions performed by the company are secure and stable by blockchain technology.

Blockchain technology has been receiving increased interest due to the technologies being decentralized, transparent, fault-tolerant, and having strong security features. Security concerns regarding data privacy are widespread. There are growing concerns about malicious attacks, third-party reliance, and privacy leakage that have caused a loss of trust and huge economic losses (Mulhim, 2022). Currently, most traditional databases rely upon human supervision and large amounts of data that are generated are hard to store in a centralized manner and extremely difficult to control centers to maintain the security of data due to malicious intrusions. Blockchain technology emerged from Bitcoin back in 2008 and quickly garnered attention from many industries as a robust technology for myriad organizational big data storage and processing issues (Guo et al., 2022). Blockchain has its own characteristics in terms of delay and functionality, based on the principles of a distributed database, miners play an essential role, both as participants and maintainers of the blockchain. This unique feature of the blockchain creates verification and transparency of the transactions. The inclusion of consensus algorithms provides stability to the blockchain system and solves the problem of data asynchrony.

Comparing Public, Private, and Consortium Blockchain

The three levels of blockchains differ in restrictions to participants, access control, degree of decentralization, security level, transaction speed, and transaction costs.

1. Public blockchain allows all users on the chain to access data that is completely decentralized and offers a high level of security, slow transaction speed, and higher transaction cost.
2. Private blockchain allows selected users on the chain to access the data that is partially decentralized and offers a low level of security, fast transaction speed, and low transaction cost.
3. Consortium blockchain allows selected users on the chain to access the data that is partially decentralized and offers a medium level of security, fast transaction speed, and medium transaction cost.

Smart Contracts

Smart contracts are based upon computer protocols that guarantee trusted transactions without the intervention of third parties. Smart contracts reduce manpower, lower costs, and optimize the automation of the system operation. Smart contracts are digital forms that allow them to integrate with computers perfectly. There is an action built into the smart contract that automatically executes when the logic is satisfied. Ethereum is one of the most famous examples of smart contract use. The Ethereum blockchain uses a simple program that runs a smart contract which is a collection of codes (functions) and data that resides on the Ethereum blockchain. While smart contracts can be fed data, they cannot be controlled by a user. Furthermore, they are deployed to the network and run as programmed to execute. Smart contracts are automatically enforced via code when the defined rules (contract logic) are met. Smart contracts depend upon information from an external database because smart contracts cannot get information about real-world events. The dependence on external information can jeopardize consensus, which is an important security feature and effective decentralization. Apart from externally sourced data that can compromise the blockchain, another limitation of smart contracts is the maximum contract size. Most current traditional systems are based on centralized databases and cannot adapt to a new distributed system. Blockchains transparent distributed ledgers can adapt traditional legacy systems to effectively manage data and resist attacks to provide a safer environment.

Application

Industries are struggling with traditional, inefficient technology that does not support new innovative ventures. For instance, there is a huge transformation in the energy industry with the inclusion of multiple distributed energy resources such as renewable energy and battery storage systems that have triggered a need to shift power distribution from a low-efficient centralized model to a decentralized distribution system (Erturk et al., 2020). Organizations are changing and growing at a phenomenal pace and are becoming increasingly crucial to the social and economic fabric of the world. The advent of Blockchain 2008 has been labeled as one of the most revolutionary technologies that are applicable in different fields (Khan and Massod, 2022). The legacy operational systems have been built around the principle of control by one central party (Kirpes et al., 2019; Siano et al., 2019). The central control principle aided in data security but also left a lot of limitations and vulnerabilities. As data technology advanced there has been a constant need for better data security that has evolved in the market but has not been sufficiently met (Khan and Masood, 2022). At the core of data security is the removal of centralized functions that remain exposed to cyber-attacks. The core structure of blockchain technology creates a robust solution to provide a next-generation technology that decentralizes control thereby creating data security and removing weaknesses that exist with centralized control (Liu et al., 2018). According to Van Cutsem et al. (2019), communities can decrease the overall cost of energy by increasing cooperation and distribution of renewable energy. However, the challenges of creating cooperation between energy companies and consumers must be resolved, which is achievable through smart contracts.

Agung and Handyani (2022) assert that proof of work is resource intensive and provides good data security for transactions through a blockchain-based database. Several works have noted that Ethereum is the most widely available and commonly used blockchain technology with a consensus algorithm that provides an open source, executable code (solidity) and works well with smart contracts but has weaknesses of proof of work that is not scalable and a slower rate of transactions (Afzal et al., 2020; Foti et al., 2019; MengelKamp et al., 2018).

Foti and Vavalis (2019) discovered that the installation of computational modules on smart computing devices can lower the block size thereby increasing the higher efficiency of blockchains. Hyperledger Fabric is another commonly used blockchain platform that provides a strong framework for permissioned networks creating secure and auditable functionality and providing a customizable modular architecture, but on the flip side, permission networks mean not fully decentralized, and new frameworks are limited to proofs of application, and trustworthiness depends on the network participants (Goranovic et al., 2019; Patsonakis et al.,

2019). Blockchain-based databases that include AI are even better suited to defend against cyberattacks. For instance, use of DeepCoin, which is a blockchain and deep learning-based connectivity system that uses blockchain for smart contacts, and uses deep learning for security and intrusion detection.

It is important to differentiate and understand the concept of private, public, and consortium-based blockchains. It is a common misunderstanding to believe that public blockchains are not very secure because they are completely decentralized. All blockchains offer decentralization due to the basic structure of the technology; however, in a public blockchain users are hidden behind a layer of cryptography while all the transactions are public (Di Silvestre et al., 2020). Although public blockchains are inherently less secure because studying multiple transactions can provide a possible match of user keys with different users. In addition, the use of public blockchain is extremely energy intensive because each transaction recorded on the blockchain must be broadcast to each member of the network. On the other hand, private blockchains minimize computational load and broadcast data only to authorized and trusted parties on the network creating more of a centralized structure (Rathor and Saxena, 2020). Consortium blockchains are a hybrid structure that provides less computational stress and only a few trusted nodes are given read/write permissions. Since the nodes are chosen in a decentralized manner there is no consolidation of power (Khajeh et al., 2020). The application of blockchain-based databases provides a sound solution to the rapidly growing industries that seek to resolve real-world issues; however, the main issue is to enable and build upon existing solutions and give non-technical people a compressive understanding of blockchain applications.

The world is investing in the development of electric vehicle (EV) technology. While Tesla is the world's largest manufacturer of EVs there are currently more than 30 EV manufacturers globally, a trend that continues to gain traction to move away from combustion engines that use crude oil (Guo et al., 2022; Zhang et al., 2022). The increasing population of EVs demands a future smart grid that establishes a fair and private electricity exchange. Electricity distribution and maintenance have been a difficult problem for smart grid applications. The use of an auction scheme seems like a viable solution to alleviate the problem of electricity distribution to smart grids. In a normal auction process, a centralized manager conducts the auction. There are two problems with the centralized auction system: it is possible to have a single point of failure and the auctioneer may be malicious. The centralized auctioneer can compile users and infringe on the interest of the acquiescent users, destroying the trust between buyers, sellers, and auctioneers. Although auctions are an easy and efficient way to distribute energy, their flaws inherently arise from a central authority that can be eliminated using blockchain technology.

Blockchain's core principles of decentralization can facilitate auctions to bring new opportunities for auction and the smart grid. Auctions can achieve an effective allocation of energy resources in a transparent and fair manner that is ensured by the properties of blockchain technology. Among the opportunities for blockchain-based technology for the smart grid is a decentralized database that is immutable, automated smart contracts that are in a decentralized environment, efficiency and transparency of transactions, and privacy protection.

Merging Blockchain and Existing Quality Management Techniques like Six Sigma

Integration of blockchain and Six Sigma needs to be thoughtfully and carefully planned and implemented as desired by an organization's specific needs and goals.

The integration of blockchain and Six Sigma brings several benefits to an organization:

1. **Enchanted transparency and traceability:** Blockchain technology facilitates a secure and transparent record. Integrating Six Sigma principles can analyze transaction data and identify areas of improvement.
2. **Improved security:** A high level of security and authentication can be achieved through the application of blockchain technology. This can prevent fraud and unauthorized access to data.
3. **Increases customer satisfaction:** Blockchain technology provides greater visibility and transparency within a transaction. The application of Six Sigma principles can identify areas where customer satisfaction can be improved.
4. **Streamlined processes:** One of the key significance of the Six Sigma methodology is to reduce defects and minimize process variability. Blockchain integration helps automate and streamline processes to drive overall operational efficiency.
5. **Data-driven decision-making:** Six Sigma and blockchain rely on data analysis to make informed decisions. Thus, the application of Six Sigma tools and blockchain records can support continuous improvement.
6. **Compliance and auditing benefits:** Blockchain underpins tamper-proof and immutability, this can be leveraged to streamline compliance processes.

Blockchain and Six Sigma are two distinct concepts that have gained significant attention in recent years. Blockchain is a digital ledger technology that enables secure, decentralized transactions, while Six Sigma is a quality management methodology that seeks to minimize defects and improve efficiency in organizational processes. While these concepts may seem unrelated at first, there are several ways in which they can be integrated to enhance business operations and achieve better results.

Blockchain technology has become a powerful tool for securing and authenticating transactions in various industries. One of the key benefits of Blockchain is its ability to create tamper-proof records that can be shared securely between parties. This feature is particularly valuable in industries that rely on trust and transparency, such as finance, healthcare, and supply chain management. By using Blockchain, organizations can track the movement of goods and services, verify the authenticity of products, and improve the overall transparency of their operations.

On the other hand, Six Sigma is a data-driven methodology that seeks to improve organizational performance by reducing defects and minimizing process variability. Six Sigma achieves this by using statistical tools and techniques to measure and analyze data, identify areas of improvement, and implement solutions to optimize processes. By applying Six Sigma principles, organizations can achieve greater efficiency, reduce costs, and enhance customer satisfaction.

The integration of Blockchain and Six Sigma can result in several benefits for organizations. One such benefit is increased transparency and traceability of transactions. Blockchain can provide a secure and transparent record of all transactions, which can be used as a basis for Six Sigma analysis. By analyzing transaction data using Six Sigma tools, organizations can identify areas of improvement and implement solutions to optimize processes.

Another benefit of combining Blockchain and Six Sigma is enhanced security. Blockchain technology provides a high degree of security and authentication, which can be used to prevent fraud and unauthorized access to data. By integrating Six Sigma principles, organizations can further enhance their security measures by identifying potential vulnerabilities and implementing solutions to mitigate risks.

Furthermore, the use of Blockchain and Six Sigma can lead to improved customer satisfaction. Blockchain technology provides customers with greater visibility and transparency in their transactions, which can enhance their trust in the organization. By applying Six Sigma principles, organizations can also identify areas where customer satisfaction can be improved and implement solutions to address these issues.

Thus, the integration of Blockchain and Six Sigma can result in several benefits for organizations. By using blockchain technology to create secure and transparent records of transactions and applying Six Sigma principles to analyze and optimize these transactions, organizations can achieve greater efficiency, security, and customer satisfaction. As these technologies continue to evolve, it is likely that we will see more examples of how they can be integrated to enhance business operations and achieve better results.

Conclusion

Blockchain technology features a public ledger that is distributed over a network and records transactions executed among network participants. Transactions on the blockchain are verified through network nodes and majority consensus before each transaction is added to the blockchain. Recorded transactions on a blockchain cannot be changed or erased. Many early adopters of blockchain technology have reported breakthroughs in many daily activities and business processes. Thereby supporting blockchain technology as a productive and viable business solution.

Many industries are seeking modernization from their legacy systems, particularly in terms of leveraging technology to improve efficiency, accessibility, and trust in financial transactions. This could include items like mobile banking, blockchain technology, and other digital solutions that can streamline transactions and reduce costs. Blockchain technology offers security and transparency that can be achieved through the use of smart contracts.

Security: Digital assets that are stored on blockchain-based databases are accessed by anyone in the world. The security of blockchain technology provides a much better solution than storing cash and cash transactions that create a burden. The safety of digital currency makes it not only safer but easier to use because every transaction that is entered gets recorded and the blockchain network verifies its authenticity.

Transparency: Blockchain technology is an open-source technology where all users on the chain have access to the stored data. For instance, auditors or regulators have the ability to review the transaction for security. The lack of one central authority makes the technology transparent for all users. Furthermore, blockchain technology provides faster and easier processing to gain trust as compared to traditional procedures. A digital, paperless system provides more transparency throughout the transaction process for all users. Human error and fraudulent employee behavior can both be avoided by ensuring that all records of transactions performed by the company are secure and stable through the use of blockchain technology.

Smart contracts: Smart contracts are based upon computer protocols that guarantee trusted transactions without the intervention of third parties. Smart contracts reduce manpower, lower costs, and optimize the automation of the system operation. Smart contracts are digital forms that allow them to integrate with computers perfectly. There is an action built into the smart contract that automatically executes when the logic is satisfied. Furthermore, they are deployed to the network and run as programmed to execute. Smart contracts are automatically enforced via code when the defined rules (contract logic) are met. Smart contracts are dependent upon information from an external database because smart contracts cannot get information about real-world events. The dependence on external information can jeopardize consensus, which is an important security feature and effective decentralization. Apart from externally sourced data that can compromise the blockchain, another limitation of smart contracts is the maximum contract size. Most current traditional systems are based on centralized databases and cannot adapt to a new distributed system. Blockchains transparent distributed ledgers can adapt traditional legacy systems to effectively manage data and resist attacks to provide a safer environment.

Six Sigma focuses on reducing variation and defects in processes to improve efficiency and quality. Therefore Six Sigma and blockchain integration can enhance business operations in the following ways:

1. **Enhanced Traceability:** Integrating Six Sigma methodologies can ensure organizations can analyze stored data on the blockchain to identify areas of inefficiency, reduce defects, and streamline processes.
2. **Improved Data Integrity:** The decentralized nature and cryptographic algorithm component of blockchain's make it highly secure and resistant to tampering. Leveraging Six Sigma methods, organizations can ensure the accuracy and reliability of stored data.
3. **Smart Contracts and Process Automation:** Blockchain technology supports smart contracts, that are self-executing contracts with predefined parameters. Use of Six Sigma methodologies in conjunction with smart contracts can automate certain processes and reduce the likelihood of human error.
4. **Supply Chain Optimization:** Both blockchain and Six Sigma play a complimentary role in managing data on a supply chain. Blockchain can enhance the transparency and trust of the data by enabling real-time tracking of goods. The application of Six Sigma can be appointed to analyze supply chain data to identify bottlenecks, optimize inventory, and reduce defects.

Chapter Summary

Since its introduction in 2008 with the launch of Bitcoin, the first of its kind of digital currency traded without relying on intermediaries in an immutable and publicly verifiable manner. Big data and technological evolution have been occurring in cycles. For instance, digital money creation, registering, confirming, and transferring contracts, and sector application.

The blockchain era represents the expansion of blockchain technology beyond the realms of cryptocurrency and finance. A crucial role played by Blockchain technology is immutability. Immutability refers to the inability to alter or tamper with data once it has been recorded on the blockchain. For instance, in sectors like the government, blockchain technology can enhance transparency, efficiency, and trust. Government agencies can employ blockchain for the secure storage and sharing of sensitive citizen data and reduce the risk of data breaches. In the healthcare industry, blockchain technology can be utilized for securely managing patient records, and similarly, in the equation sector, blockchain can be used to verify and authenticate education credentials and certifications to prevent fraud and ensure the validity of qualifications. Blockchain has significant potential applications in supply chain management where the technology can enhance transparency and traceability by allowing stakeholders to track the movement of goods across the supply chain.

This chapter thoughtfully concluded on important advantages of blockchain technology which are as follows:

1. Constructing a shared repository of data that is maintained by peers (no third-party intermediary). All users have access to the data and viewing transactions

and storing information on nodes prevents data loss in case of unanticipated events.

2. Embracing trust among users through digital signatures and validations ensures every node and user behaves ethically.
3. Building potential infrastructure that transforms the next digital fabric of the world by creating a worldwide repository of data that can potentially be accessed by everyone.
4. Creating transparency and guaranteeing that everyone on the platform can read the final state of the transactions and have full access to the history of the transaction.
5. Automation with infusing smart contracts.
6. Immutability ensures that data cannot be erased or changed.
7. Decentralization creates consensus and removes one central authority.

In addition, the chapter delves into the disadvantages of blockchain technology which are as follows:

1. High power consumption is attributed to the algorithms built in for mining that require expensive hardware. A massive amount of computation is required for data mining.
2. As a current copy of the blockchain is available to all users, the data replication requires more space.
3. Immutability and transparency eliminate users who do not meet the standards, while a positive for the industry and a negative for the user who had the potential to change business practices but can no longer participate because of some negative data.
4. Smart contracts must rely upon external data—this process needs to be either streamlined or the data needs to be injected into the blockchain. A lack of data governance can cause strong reputational issues and become the weakest part of the process.

There are three different types of blockchain infrastructure, and the chapter offers and comparison of the three types of blockchain: public, private, and consortium blockchain. The three levels of blockchains differ in restrictions to participants, access control, degree of decentralization, security level, transaction speed, and transaction costs.

1. Public blockchain allows all users on the chain to access the data that is completely decentralized, and offers a high level of security, slow transaction speed, and higher transaction cost.
2. Private blockchain allows selected users on the chain to access the data that is partially decentralized, and offers a low level of security, fast transaction speed, and low transaction cost.
3. Consortium blockchain allows selected users on the chain to access data that is partially decentralized, and offers a medium level of security, fast transaction speed, and medium transaction cost.

Take Away Lessons from the Chapter

1. Each transaction on the blockchain is verified through network nodes and majority consensus before it is added to the blockchain.
2. Blockchain's decentralization function ensures that no single entity has control over the network or the transactions that take place on it, making it more resistant to fraud, censorship, or attacks.
3. There are currently four realized periods in blockchain evolution. Blockchain 1.0 Era (digital money creation), blockchain 2.0 Era (registering, confirming, and transferring contracts), blockchain 3.0 Era (sector application), and blockchain 4.0 Era (beyond evolution to adoption).
4. The safety of digital currency makes it not only safer but easier to use because every transaction that is entered gets recorded and the blockchain network verifies its authenticity.
5. It is a common misunderstanding to believe that public blockchains are not very secure because they are completely decentralized.
6. Blockchain is a digital ledger technology that enables secure, decentralized transactions, while Six Sigma is a quality management methodology that seeks to minimize defects and improve efficiency in organizational processes.
7. By using blockchain technology to create secure and transparent records of transactions and applying Six Sigma principles to analyze and optimize these transactions, organizations can achieve greater efficiency, security, and customer satisfaction.

Chapter Exercises

1. What are some of the opportunities of blockchain technology? Why do you think it is important for organizations to understand the benefits of blockchain technology?
2. Break into teams and discuss why business leaders should consider blockchain technology. Create a list of pros and cons and discuss the importance of them.
3. How do you think business tools such as Six Sigma fit into blockchain technology and how can organizations benefit from it?

Blockchain and Exploitation

Ledgers have been around for thousands of years to keep track of transactions. As societies develop, so does the use of ledgers. Today, ledgers have evolved into digital and computerized formats. At the basic level, blockchain technology is a disruptive new platform technology that enables the improved ability for verification and recording of exchange of value among an interconnected set of users. Blockchain technology offers a secure and transparent way to track the ownership of assets throughout the transaction process: before, during, and after transactions are recorded. The name blockchain emanates from the format of the technology. Every transaction is recorded on a block and a set of transactions across the entire platform is the chain, therefore blockchain. One of the biggest challenges faced by law enforcement agencies is the anonymous nature of digital currency transactions. The lack of transparency makes it difficult for law enforcement agencies to trace the movement of funds and identify the individuals involved in illegal activities. As a novel technology, regulatory and governance considerations must be thoughtfully incorporated into blockchain-based networks to truly experience the potential of the technology. The security of physical assets is a major concern for organizations. Removing human intervention from the process creates an environment that prevents error and deception. Blockchain has been a rapidly evolving Internet database technology that has been implemented in several industries to gain efficiency.

Introduction

Ledgers have been an integral part of human history, whose use can be traced back more than 7,000 years ago (Friedlob and Plewa, 1996). Mesopotamia recorded simple sales and purchases of goods on clay tablets. Counting was even impacted by the development of ledger keeping in antiquity. Record maintenance of transactions seems to be a mutual in human life. Humankind has progressed a long way since clay tablets, and now modern-day computing has transformed financial ledgers into super-fast, automated, and precision relational databases. The rapid evolution of technology is the center of cybersecurity and transparency concerning digital transactions. Blockchain technology has the potential to provide the next generation of innovative improvements to modernize the need for ledger record keeping (Allayannis et al., 2018). There have been several implementations and upgrades surrounding the digitization of records; however, the widespread adoption of blockchain technology has been slowly moving particularly due to a lack of understanding of the technology and regulatory opacity. Many industry leaders admit that blockchain technology is indeed the new wave of designing digital asset repositories; however, admit they have little to no knowledge of how the technology works.

What is Blockchain?

At the basic level, blockchain technology is a disruptive new platform technology that enables the improved ability for verification and recording of exchange of value among an interconnected set of users. Blockchain technology offers a secure and transparent way to track the ownership of assets throughout the transaction process: before, during, and after transactions are recorded. The name blockchain derives from the format of the technology. Every transaction is recorded on a block and a set of transactions across the entire platform is the chain, therefore, blockchain. Blockchain technology is versatile and enables any network of users to track and trade anything of value: digital currency, digital assets, transactions, or any other data that is important to the users.

Blockchain's basic functions include:

1. **Consistency:** Blockchain records maintain historical data on all transactions, and all users on the blockchain have identical copies of the records.
2. **Democratic:** The network itself has built-in agreements and rules that govern the system. Any changes made to transactions are democratically approved by the authorized user; there is no centralized authority.
3. **Secure and accurate:** The technology is based on cryptography that protects all data and information by using digital keys and signatures to access the data in the ledger.
4. **Segment and private:** The use of digital keys and signatures and predefined rules allows users to access the entire ledger network.
5. **Permanent and tamper resistant:** The technology is not based upon one single centralized point of control; therefore, the details of all the recorded transactions cannot be altered without the interaction of the entire network.
6. **Rapidly updated.** Blockchain ledger features real-time user information as any changes are made on the ledger, and every user copy is swiftly updated to reflect the changes on the network for all users.
7. **Intelligent:** Blockchain technology facilitates the use of smart contracts to automatically execute predefined contracts to create time and cost efficiencies.

The rise of digital currencies has created significant challenges for law enforcement agencies worldwide. While digital currencies offer numerous benefits, such as decentralization and anonymity, they have also been exploited by criminal organizations for illicit activities such as money laundering, drug trafficking, and terrorism financing. As a result, governments and regulatory bodies have been struggling to keep up with the pace of technological change and have been slow in creating effective measures to counter digital currency-related crimes.

One of the biggest challenges faced by law enforcement agencies is the anonymous nature of digital currency transactions. The lack of transparency makes it difficult for law enforcement agencies to trace the movement of funds and identify the individuals involved in illegal activities. Additionally, digital currencies are decentralized, meaning that they are not subject to the same regulations as traditional financial institutions. This makes it easier for criminals to exploit the system without fear of being caught.

Using of the dark web to conduct illegal activities is another challenge facing global organizations. The dark web provides a platform for criminals to operate anonymously and evade law enforcement agencies. Criminals can use digital currencies to purchase illegal goods and services on the dark web, making it challenging for law enforcement agencies to monitor and investigate these transactions. To address these challenges, governments and regulatory bodies are exploring various options. One approach is to increase regulations and oversight of digital currencies. Some governments have implemented strict regulations on digital currency exchanges and wallets to reduce the risks of money laundering and terrorist financing.

Additionally, some countries have banned the use of digital currencies altogether, while others have introduced digital currency-related legislation. An approach is to improve international cooperation and collaboration among law enforcement agencies. As digital currencies are borderless, it is essential to have international cooperation to combat their illegal use. The creation of international task forces and the sharing of intelligence could help to reduce the risks of digital currency-related crimes. Therefore, digital currencies offer numerous benefits, but they also pose significant challenges for law enforcement agencies. The anonymous and decentralized nature of digital currencies has made it easier for criminals to exploit the system. Governments and regulatory bodies need to continue exploring various approaches to reduce the risks of digital currency-related crimes and ensure that these currencies are not used to facilitate illegal activities.

Historically blockchain technology has been widely used in clearing payments and settlement functions in the financial services industry, creating and using of digital assets such as cryptocurrencies, digital identity within enterprises, and incorporating smart contracts to streamline business processes. Digital identity is the creation of a secure, private, and tamper-resistant single source of validated data on each customer within an enterprise. Digital identities are a specialized feature of blockchain technology that creates time and cost savings during the customer onboarding process. Global industries have consumer safety at the heart of their operations. When customers are onboarding, there is a plethora of personal information that is verified to meet the compliance requirements of the regulations. KYC is an important financial metric that provides valuable information regarding the client's risk profile, suitability, and buying habits to protect the client from abusive financial practices. Onboarding clients is no easy feat, it is an extremely cumbersome process for both the client and the organization. Incorporating blockchain technology that facilitates the creation of a digital identity for the client moves the client onboarding process to new heights by removing wait time, research, filling out long forms, sending documents back and forth for signature, and funding accounts. All these functions are very prevalent in many industries, therefore, understanding the benefits of blockchain technology is extremely crucial to leaders who have the power to pivot their organization's future success to gain a competitive advantage in their industry. The benefits of blockchain technology help create efficiencies by constructing digital identities that are consistent, secure, accurate, private, tamper-resistant, and rapidly updated on the ledger across the organization. Digital identities created on the blockchain are consistent as the digital record is the only single source of verity and is shared with the entire organization. All authorized users on the blockchain platform have digital keys to securely, accurately, and privately access the data. A consensus is required from most of the network users to change records, thus, the technology is tampered-resistant. All users who have access to the customer's digital identity receive real-time updates as changes are made to the customer records.

Blockchain technology creates tremendous efficacy to streamline business processes using smart contracts. Smart contracts are automated contracts that have predefined functions. Smart contracts are automatically executed based on logic (if then). For instance, when we receive payment, then execute shipping of the product. The smart contracts are stored on the blockchain which ensures that there is an audit trail of events and that contract fulfillment terms are fully executed. The use of smart contracts has many possible advantages and applications: client onboarding, trade order generation, regulatory reporting, clearing, and settling of financial transactions, cross-border payments, compliance reporting, real asset transactions such as real estate, and data storage. Many areas where blockchain technology can potentially create value are customer loyalty programs, medical recordkeeping, supply chain management, passport and customs controls, tax collections and payments, customer payments, government benefits distribution, and inventory controls. Blockchain technology has already impacted the financial services industry in the following areas: intrabank cross-border payments, interbank cross-border transactions, cross-border payments, cross-border remittances, corporate payments, and person-to-person transactions. The incorporation of blockchain technology in various cross-border and peer-to-peer transactions has lowered transaction costs and administrative costs created shorter settlement times and aided in fewer human errors and exceptions. The need for money exchange and flows has evolved tremendously. Blockchain technology can assist financial institutions to achieve cost and efficiency savings when sending and receiving payments across borders.

Global financial institutions must go through a rigorous intermediary process that is relatively slow to move due to regulations, and fragmented, and limited technology functions. In addition to traditional constraints, international transfers have high fees and the tracking mechanism for payments is very uncertain. Global financial organizations can gain many efficiencies by adopting blockchain technology to execute cross-border financial transactions. An enterprise blockchain can facilitate real-time cross-border payment, eliminate the use of an intermediary and substantially reduce costs to a negligible fee. In addition to cost and efficiency savings, the transaction would be secure, private, and validated. Although the masses do not disagree with the potential benefits of blockchain-driven products, one of the main arguments against blockchain technology is the newness of technology that does not have many years of developed platform information to back it up. There have been proven relational databases that have been on the market for several decades and resistance to new technology due to the unknown is certainly one perspective that leaders should consider regarding blockchain technology and its various potential advantages for the future of their firm.

Deeper dive into blockchain technology reveals that it is a network-effect technology. Network effect technology is based upon the premise that more users will use it if anything becomes more valuable to its users. Several types of networks that correspond to their valuation concept and relation to data. A basic network is broadcast. For instance, when a group of people participates in the same broadcast from a single source. Broadcast is information flows in one direction none of the participants relate to one another the information flow is unidirectional. Another type of network is a homogenous network where users use the platform for similar purposes such as Skype. While Skype can encourage conversations between two individuals, the technology is also capable of boarding many users to make phone calls, texts, and videoconferencing to make the network more valuable. Conversely, a heterogeneous network is where diverse sets of users use a platform for varying purposes. For instance, LinkedIn where a large user base and many self-informing groups participate on the network. The value of the LinkedIn network grows as more and more people participate in the network.

As a novel technology, regulatory and governance considerations must be thoughtfully incorporated into blockchain-based networks to truly experience the potential of the technology. The core feature of blockchains is a decentralized, open-source technology that does not follow one owner or authorized approver. Therefore, from a regulatory perspective, establishing a governance structure to monitor the reliability and accountability of public blockchains is critical. As regulators decide upon proper regulation of the technology, industry leaders must do the same. Blockchain technology evolution has made it the next most promising technology with valuable product and service applications. Business leaders must consider the next wave of business evolution for their organization to pursue the opportunities of this new technology. If leaders choose to wait, they may forgo the next generation of technology architecture. Industry leaders in complicated industries such as global supply chain management experience a high degree of intermediaries that require a higher level of data security for transactions.

Structure

Blockchain integration and the Internet will lead to historic changes in the industry that creates uncertainty in the future (Christidis and Devetsikiotis, 2016). Blockchain has been a rapidly evolving Internet database technology that has been implemented in several industries to gain efficiency. Blockchain technology divides data into many blocks that are passed through in secrecy and digitally validated. Blockchain technology connects blocks to a data network that is essentially different from traditional databases (Wang, 2022). The data stored in the blockchain is secured and shared, so it can be distributed in the data library. Blockchain is a naturally decentralized distributed storage technology that enables the establishment of trust relationships without a network (Xiao, 2020). Blockchains are uninterrupted digital transactions that can be programmed to record anything of value and the decentralized database containing data is modifiable and managed by a cluster of computers without one central authority. The distributed ledger technology of the blockchain has a security record for every transaction on the Internet. Irreversibility is the immutability or non-tampering mechanism of the blockchain. Although it is theoretically possible to reverse a transaction, 51% of the computing power is needed to recognize the reversal. Anti-censorship is the blockchain's cornerstone principle to keep a record of every single transaction made to prevent the tracking of blockchain projects. Near real-time settlement is the blockchains internal updating time, which ensures that the transaction confirmation time is almost the same and reduces data insecurity.

Advantages of Blockchain

Blockchain-based models create distributed data storage and point-to-point transmission consensus mechanisms with encrypted algorithms. The five core advantages of blockchain technology are decentralization, openness, autonomy, tamper resistance, and anonymity.

Decentralization

Blockchains offer decentralized accounting and storage without centralized management. The nodes on a blockchain have equal rights and the system maintenance is handled by specialized maintenance nodes. Decentralization promotes transparency and diversification. Blockchain technology constructs transparency and diversification to adapt markets to their natural structure.

Openness

Blockchains facilitate an open platform. Anyone who has access to the blockchain has access to the data in near real-time. Only specific information is hidden; however, anyone with access to the platform can query the data through a secure interface. Resource sharing is at the heart of big data networks. While big data networks are very convenient for data sharing, they lack robust security because not only all network users can access server resources, but information and data are shared between different terminal devices. Sharing of data in this manner creates data security vulnerabilities.

Autonomy

Blockchain's autonomy is dependent upon specifications and protocols. Data is freely converted from nodes in the system. In a blockchain architecture, each node has its own smart contract data that adheres to a predefined set of terms, such as customer information, dynamic contracts, etc. that can track transaction information among users. Blockchains can automatically generate smart contacts, collect, and store distributed transactions, centrally managed distributed transactions, and realize transactions.

Tamper Resistant

Blockchain data cannot be tampered with. Once the transaction and user information is added to the blockchain, it cannot be modified. The stability and reliability of the blockchain are highly regarded because more than half the nodes must be manipulated to modify the content, and this takes enormous computing power. Hackers have been known to use their proficiency with computers to attack network systems and implant viruses. In addition, data transfer impacts are also problematic. Information leakage can also occur through communication lines, especially with newer technology. Illegal users monitoring communication lines to obtain personal information about individuals and organizations have not been very successful in shielding and securitizing consumer data.

Anonymous

A fixed algorithm built into the blockchain facilitates exchanges between nodes. Therefore, users do not have to disclose their identities. This is an important feature to build trust. Users have changed, attacked, and destroyed network systems autonomously, which leads to information leakage and network paralysis.

Application

Blockchain application poses questions such as the high cost of implementation, organizational readiness, and governance.

High Cost of Implementation

Several authors have pointed out the challenging nature of the technology and that the technology has a high cost of implementation (Casey and Wong, 2017; Wang et al., 2019; Janssen et al., 2020). In addition, blockchain adoption may be adversely impacted by the organization's resistance to moving to a new expensive technology while preexisting legacy systems are functional and cheaper to operate. Furthermore, blockchain standardizing to create the flow of information exchange is yet another factor that adds to the high cost of implementation in the supply chain industry.

Organizational Readiness

Organizational readiness relates to an organization's willingness to spend financial and technical resources to implement new technology. The integration of blockchain technology directly depends upon investment in software and hardware technology that uses sophisticated information systems for collecting, storing, and communicating data (Iansiti and Lakhani, 2017). According to Kouhizadeh et al. (2021), some important factors that have created hindrances to blockchain adoption are lack of management commitment and support, lack of product knowledge, lack of collaboration, synchronization, and information disclosure between blockchain supply chain members, and lack of industry involvement in standardized policies. Top leadership support that signals organizational readiness is a significant determinant of blockchain adoption and larger organizations are more likely to adopt blockchain technology than smaller and medium-sized enterprises due to the higher cost and complexity of the technology specific to supply chain management application (Clohessy and Acton, 2109; Wong et al., 2020). Blockchain technology poses an understanding of a complex technology that is yet another reason that organizations struggle in terms of adoption (Crosby et al., 2016; Iansiti and Lakhani, 2017).

Governance

Standardized governance that provides clear rules for blockchain connectivity in the global supply chain structure provides interoperability between two or more chains. Governance involves the various boundaries and conditions for using the blockchain solution. The firm's internal business processes system relies upon the technical capability and ability of the blockchain. Different jurisdictions have different regulatory conditions that need to be satisfied for compliance. Food safety globally differs in terms of customer-specific regulations. The need for new regulations and the constant evolution of the governance framework of blockchain needs exchange processes that are standardized to be effectively operable.

Organizations stand to gain supply chain financing, contracting, and participating in global business if their inventory, information, and financial flows are shared among firms using blockchain technology. Information asymmetry plays a significant role in financing because the banks that provide the working capital do not have quality information on their clients, for instance, the quality of the assets and the liabilities. Short-term loans are a common practice to procure financing for working capital needs for organizations. Organizations may have contracts with several banking institutions for the same asset or get financing for one reason and use the funds for another purpose. These are all loopholes that create control risk and lending risks for banking institutions. Conventionally banks' lending processes are designed to control risks of information asymmetry, but these controls increase the transaction cost, slow the lending process down and reduce capital available for small firms to borrow.

Accounting functions are another area where blockchain technology can provide significant improvements. Accounts payable management involves many processes such as invoicing, reconciling against purchase orders, tracking payment terms, and review of data and approval at each step. ERP has been successful in automating some functions in accounting but there remains a considerable amount of manual intervention. Due to the heavy manual nature of the data, there are often errors and conflicts.

Counterfeit

In many industries such as food, healthcare, finance, etc., there is huge concern about counterfeiting measures. Blockchain technology offers tractability of the data directly and provides anti-counterfeiting measures. Reduction in manual processes, physical documents, intermediaries, multiple checks, and verifications on the port of entry and exit are areas of improvement that are currently slow, costly, and riddled by low visibility about the shipment records. Connecting inventory flows, information flows, and financial flows and sharing them with all transactional parties enables organizations that use blockchain technology to easily reconcile purchase orders and invoices, and process and track payment for goods and services. The process would be divided into following steps:

1. The supplier receives an order.
2. A bank that is on the same blockchain as the supplier can immediately process the working capital to the supplier.
3. After the delivery of the merchandise to the buyer.
4. The bank can promptly obtain payments.

The smart contracts that are built into the fabric of the blockchain provide conditions for all participants on the blockchain to automatically reconcile transactions and remove conflict between the bank and the borrowing firm due to increased visibility and authentication of the data that is already performed on the blockchain.

The security of physical assets is a major concern for organizations. Removing human intervention from the process creates an environment that prevents error and deception. For instance, contaminated or counterfeit products might be tagged and then introduced into the supply chain, either by human error or by unscrupulous entities on purpose.

Du et al. (2019) believe that blockchain technology can considerably improve supply chains from end-to-end traceability, speed, coordination, and financing. Blockchain technology provides a powerful solution to address the deficiency in the traditional supply chain. While blockchain technology is a promising resolution, blockchain platforms must build an ecosystem based on industry-wide consensus and commitment toward standardization.

Conclusion

Ledgers have been used for thousands of years to keep track of transactions and important information. The earliest known records of ledgers date back to around 3200 BCE in ancient Mesopotamia, where clay tablets were used to keep track of goods and commodities. The ancient Egyptians also used papyrus scrolls to record transactions and other important information. As societies developed, so did the use of ledgers. In medieval Europe, monks kept records of land ownership and other important information in large bound books called "cartularies." By the 19th century, with the rise of modern banking and commerce, ledgers became even more important and were used to record transactions in businesses, government agencies, and other organizations. Today, ledgers have evolved to become digital and computerized, with the widespread use of accounting software and other digital tools to track financial transactions and other important information. Despite the changes in technology, the basic purpose of a ledger remains the same: to keep track of important information in a structured and organized way.

Implementing a blockchain application can pose various challenges, including:

- **High cost of implementation:** Developing and deploying a blockchain application can be expensive due to the need for specialized skills, hardware, and software. Additionally, the cost of running and maintaining the network can be high.
- **Organizational readiness:** Implementing a blockchain application requires significant changes to existing business processes, which may require significant changes to the organization's culture and mindset. There may be resistance to change or a lack of understanding of the technology, which can hinder adoption.
- **Governance:** Blockchain networks are typically decentralized, which means that there is no central authority overseeing the network. As a result, governance of the network must be designed and implemented to ensure that all participants follow the rules and regulations of the network.
- **Scalability:** As the size of the network grows, the number of transactions that can be processed on the blockchain may become limited, leading to slower processing times and higher fees.
- **Security:** Although blockchain technology is considered secure, it is still susceptible to certain types of attacks, such as 51% attacks. Additionally,

the security of the network can be compromised if there are vulnerabilities in the code or if participants fail to follow security best practices.

- **Interoperability:** Different blockchain networks may use different protocols and standards, which can make it difficult to transfer data and assets between different networks. This can limit the usefulness of blockchain technology and hinder adoption.

Chapter Summary

Mankind has progressed a long way since clay tablets, and now modern-day computing has transformed financial ledgers into super-fast, automated, and precision relational databases. The rapid evolution of technology is at the center of cybersecurity and transparency concerning digital transactions. There have been several implementations and upgrades surrounding the digitization of records; however, the widespread adoption of blockchain technology has been slowly moving because of a lack of understanding of the technology and regulatory opacity. Many industry leaders admit that blockchain technology is indeed the new wave of designing digital asset repositories; however, admit they have little to no knowledge of how the technology works.

Blockchain technology is a disruptive new platform technology that enables the improved ability for verification and recording of the exchange of value among an interconnected set of users. Blockchain technology offers a secure and transparent way to track the ownership of assets throughout the transaction process: before, during, and after transactions are recorded. The name blockchain derives from the format of the technology. Every transaction is recorded on a block and a set of transactions across the entire platform is the chain therefore blockchain.

Blockchain technology can be implemented in various industries, such as finance, supply chain management, healthcare, voting systems, and more. However, the development of a network infrastructure, consensus mechanism, and smart contract must be developed per industry requirements.

Advantages of Blockchain include transparency, security, efficiency and cost savings, decentralization, and trust and disintermediation. Disadvantages of blockchain include scalability, energy consumption, regulatory and legal challenges, privacy concerns, and immutability challenges.

Therefore, industries need to understand the impact of the advantages and disadvantages of blockchain technology and access if the technology is suitable and beneficial to their specific case use and implementation approach.

Historically blockchain technology has been widely used in clearing payments and settlement functions in the financial services industry, creating and use of digital assets such as cryptocurrencies, digital identity within enterprises, and incorporating smart contracts to streamline business processes. Digital identity is the creation of a secure, private, and tamper-resistant single source of validated data on each customer within an enterprise.

As a novel technology, regulatory and governance considerations must be thoughtfully incorporated into blockchain-based networks to truly experience the potential of the technology. The core feature of blockchains is a decentralized, open-source technology that does not follow one owner or authorized approver. Therefore, from a regulatory perspective establishing a governance structure to monitor the reliability and accountability of public blockchains is critical. Blockchain has been a rapidly evolving Internet database technology that has been implemented in several industries to gain efficiency. Blockchain technology divides data into many blocks that are passed through in secrecy and digitally validated.

Blockchain-based models create distributed data storage and point-to-point transmission consensus mechanisms with encrypted algorithms. The five core advantages of blockchain technology are decentralization, openness, autonomy, tamper resistance, and anonymity. Ultimately, blockchain application poses questions such as the high cost of implementation, organizational readiness, and governance.

Take Away Lessons from the Chapter

1. The rapid evolution of technology is at the center of cybersecurity and transparency concerning digital transactions. Blockchain technology has the potential to provide the next generation of innovative improvements to modernize the need for ledger record keeping.
2. Blockchain technology is a disruptive new platform technology that enables the improved ability for verification and recording of the exchange of value among an interconnected set of users.
3. The rise of digital currencies has created significant challenges for law enforcement agencies worldwide. While digital currencies offer numerous benefits, such as decentralization and anonymity, they have also been exploited by criminal organizations for illicit activities such as money laundering, drug trafficking, and terrorism financing.
4. Blockchains are uninterrupted digital transactions that can be programmed to record anything of value and the decentralized database containing data is modifiable and managed by a cluster of computers without one central authority.
5. Standardized governance provides clear rules for blockchain connectivity in the global supply chain structure provides interoperability between two or more different chains. Governance involves the various boundaries and conditions for using the blockchain solution.
6. Blockchain application poses questions such as the high cost of implementation, organizational readiness, and governance that must be evaluated in detail.
7. Blockchain technology provides a powerful solution to address the deficiency in the traditional supply chain. While blockchain technology is a promising

resolution, blockchain platforms need to build an ecosystem based on industry-wide consensus and commitment toward standardization.

Chapter Exercises

1. What are some of the important concerns surrounding the lack of enforcement of digital assets? What are some solutions?
2. As technology modernization continues, what are some important considerations for firms in terms of standardization to keep globalization in mind?
3. What are some of the important functions of blockchain and how does it connect users to create efficiencies?

Blockchain Merging of Business and Technology

Blockchain application supports an open architecture and provides improved transparency and traceability of information in near real time to all users. Blockchain characteristics include traceability, transparency, time stamped, secured, consensus driven, and censorship resistant. The potential value drivers of blockchain technology are transparency and reduced administrative costs, lowered risk of fraud, and improved control of subcontracted manufacturing. The scope of blockchain technology includes control of data and sharing of data with all stakeholders on the blockchain. Primary benefits of blockchain technology are its ability to provide a secure and transparent way to manage and transfer data. Beyond data control and sharing, blockchain technology can also enable the automation of complex business processes through smart contracts. Technology provides a way to establish trust among parties without the need for intermediaries. Some intangible benefits of blockchain are transparency, credibility, risk mitigation, and stakeholder engagement. The tangible and intangible benefits of blockchain technology can transform business solutions in practically all industries. Blockchain technology can simplify various aspects of human life, including those in the sectors you mentioned such as energy, agriculture, health, construction, manufacturing, and supply chain.

Introduction

Why would organizations need blockchain for their supply chain and beyond? Blockchain technology provides improved transparency and traceability of material throughout the supply chain and creates an open architecture for sharing information in near real time for all users. Blockchain is of interest because consumers demand more transparency, and the increasing complexity of supply chains, and it is an effective and inexpensive means for tracing materials used in the final product that aids in building confidence with increasingly environmentally and socially conscious consumers. Organizations can improve their supply chain by increasing visibility throughout the entire supply chain, decreasing administrative costs, and gaining authentication against counterfeit products. However, since the technology is still in its early trial stages in the supply chain industry, more information about data security, costs, and implications is desired to convince all stakeholders to adopt blockchain. Figure 9 shows the characteristics, potential value drivers, and scope of blockchain.

Characteristics	Potential value drivers	Scope
Traceable, transparency, time stamped, secured, consensus driven, censorship resistant	Transparency and traceability reduced administrative costs	Control of data
Immutability, distributed ledger, real-time capabilities	Lowered risk of fraud improved control of subcontracted manufacturing	Sharing of data with all stakeholders on the blokchain

Figure 9: Blockchain characteristics, value drivers, and scope
(Source: Author's own work)

Blockchain Overview

The technology functions and characteristics of blockchain are that it is a digitally distributed ledger database of records, transactions, or executed contracts that are shared across the participating users. Each transaction is timestamped and verified using a consensus algorithm through the participation of the majority of the participants on the system. Blockchain structure operates in a block format. A block of data is a storage mechanism that stores all data regarding the transaction, such as parties involved, amount, important dates of execution, etc. Every block on the blockchain is interlinked with each other; therefore, in a blockchain database users can access all previous blocks that are linked since the database retains the complete history of all data stored on the chain. This feature of blockchain is very attractive because it facilitates authentication by comparing copies of the blockchain with all users to maintain its validity. As new users add data to the blocks of the blockchain, the information is verified by the majority of the blockchain participants before it becomes a part of the shared blockchain database. Data integrity and transparency are preserved through the distributed verification method. This technology builds trust among users meaning that parties do not need to trust each other to engage in exchanges. Since blockchain is a decentralized ledger technology it is not controlled by one central point, and this eliminates failure because all participants on the blockchain will have the most updated copy of the entire database.

Logical statements can be embedded in blockchain to automate transactions and linked with smart contracts to create efficiencies in a blockchain by reducing operating costs (Nacci et al., 2015). For example, if the agreement terms are not met, using logical if-then statements, the transfer penalty fee smart contract is automatically executed. Blockchain technology has been very appealing to the financial industry. Many early adopters such as Visa, RBS, and Wall Street firms have shown their willingness to adopt the technology. Blockchain technology keeps gaining momentum in several industries, where industry leaders have realized that the technology can be customized. In 2017, Deloitte a survey revealed that 35 percent of organizations surveyed are aggressively pursuing blockchain as a solution. Some industry pioneers are Skuchain, this organization is focused on B2B trade and supply chain to target the global finance market that comprises of several entities including buyers, sellers, logistics providers, banks, regulators, and third parties. Blockchain implementation in the supply chain provides users with functions to record price, date, location, quality, certification, and any other relevant information regarding the transaction to effectively manage the supply chain.

Primary benefits of blockchain solutions in the supply chain include traceability, counterfeit measure, improved visibility, and reduced administrative costs.

1. Increased traceability in the supply chain from sourced materials to the final product, up to and including distribution. Higher traceability ensures food safety, provides confidence to the consumer, and creates credibility, and public trust.
2. Minimizing counterfeiting efforts to gain credibility in the industry.
3. Increased visibility over outsourced contracts in manufacturing to ensure a higher level of compliance.
4. Assist in achieving operational efficiencies by reducing paperwork and administrative costs.

There are several important value drivers for blockchain technology to translate into a robust solution for the supply chain industry. For instance, food transparency is in high demand, and all parties involved in the food industry are continuously seeking better ways of tracking. The incorporation of blockchain technology in the food industry supply chain provides more transparency and accuracy from end-to-end tracking. Socially conscious consumers are willing to pay higher prices for sustainably and ethically sourced products. It is estimated that more than 55 percent of consumers are willing to pay a premium price for companies promoting social responsibility. In addition to providing traceability in the food supply chain, blockchain can digitize physical assets and create a decentralized, immutable record of all transactions that build trust through validity and creates transparency about the product.

Counterfeit and subpar products that can cause serious illness can be weeded out through the application of blockchain technology. For instance, there is a big problem with prescription drugs that are counterfeited. It is estimated that 10–30 percent of drugs sold in the open market are counterfeit. Pharmaceutical companies and regulators have not achieved much success in resolving this serious and immediate problem that leads to hundreds of thousands of deaths and costs firms billions of dollars in monetary damage. A blockchain-based supply chain allows detailed tracking of each component. Raw materials used in the manufacturing of a drug can be effectively traced back to the unscrupulous supplier or even the subcontractor of a supplier who might be changing the ingredients. Visibility of detailed information such as what ingredients are being used in a particular drug helps reduce and potentially eliminate the impact of counterfeit products.

Blockchain technology offers intangible benefits such as transparency, credibility, risk mitigation, and stakholder engagement. Blockchain strengthens corporate reputation by providing real-time data and creating transparency for all users. In addition, blockchain established credibility and public trust by sharing validated and trustworthy data. Blockchain reduces the potential of reputational risk by providing counterfeit measures and helps engage stakeholders by providing them the details about a product's journey. In our media frenzy world, which reports on various social media platforms to electrify the news content, one negative post about an organization can have detrimental effects on the organization's reputation. Therefore, a post on unapproved contracting or labor disputes can result in a public relations nightmare. A proactive approach to the supply chain is desperately needed now more than ever to get ahead of the curve and galvanize blockchain features such as traceability and transparency to avoid reputational losses altogether. Quality well documented and validated data provides a quality check and builds authenticity for the data that is shared between all supply chain partners. However, achieving blockchain integration is not an easy feat. All stakeholders must embrace the technology for successful integration. Blockchain solutions offer a breakthrough potential for organizations to achieve solutions that are not possible conventionally. Adoptions considerations include traceability needs, material characteristics, and production profile, supply chain layers and partners, technology environment, and regulations.

Structure

Alkhateeb et al. (2022) found that several sectors such as energy, agriculture, health, construction, manufacturing, and supply chain are adopting blockchain technology. Technology evolution is focused on the simplification of human life. As such, technology seeks efficacy, speed, and robust frameworks that provide feasible solutions. By leveraging blockchain, organizations can streamline operations, enhance data integrity and security, and enable efficient and transparent transactions. For instance, in the energy section, blockchain can facilitate peer-to-peer energy trading, optimize chain management, and enable the tracking of renewable energy credits. The integration of blockchain technology in sectors aligns with the goal of simplifying human life by harnessing technology to achieve greater efficiency and practical solutions. Nakamoto who created Bitcoin sensationalized bitcoin; however, the underlying technology of blockchain is what introduced distributed ledger technology.

Blockchain's decentralized premise create a secure, immutable, and verifiable network for peer-to-peer transactions that are quick, safe, and transparent. The number of smart devices is estimated to increase from 16.44 billion in 2025 to 25.44 billion in 2030. The increase in connectivity of smart devices creates an interesting future for blockchain technology applications. For instance, in finance, the use of hybrid blockchain technology allows customized solutions for organizations because users can decide which transactions are made public and which are not. Although several factors of hybrid blockchain are identified currently there is no state-of-the-art use of blockchain platforms. Alkhateeb et al. (2022) evaluated 38 studies on hybrid blockchain technology and reported that data security was the top reason for organizations to implement this technology. In addition, some other key areas were to achieve transparency, and trust, improve efficiency, increase privacy, and improve quality of service.

Challenges and Solutions

1. Computationally intense protocols of blockchain technology make it very difficult for modern industrial machines to adopt the technology.
2. Replacement of the current legacy system is time-consuming and costly.
3. Older and newer technologies use different operating systems, which creates hindrances in the adoption of blockchain.
4. The consensus mechanism requires high computational power which is costly and time-consuming.
5. Technical limitations hinder traditional blockchain from scale.

Application

Potter's Five Forces (Porter, 1979) is a framework used to analyze the competitiveness of an industry or market (Fig. 10). Let's explore how this framework applies to the blockchain industry.

Threat of New Entrants: The blockchain industry is open to new entrants due to the open-source nature of the technology. This has led to the creation of many new blockchain networks and platforms. However, the high level of competition and the need for significant resources to develop a viable network can be a barrier to entry for new players. Blockchain networks require many nodes to operate effectively, and it can be challenging for new entrants to attract enough nodes to achieve the necessary level of decentralization. This can lead to a lack of trust in the network, making it difficult to attract users and developers.

Another challenge for new entrants is the dominance of established blockchain networks such as Bitcoin and Ethereum. These networks have large user bases and developer communities, making it difficult for new entrants to compete for market share. This can lead to fragmentation in the industry, with many small networks struggling to gain traction. However, new entrants can still succeed in the blockchain industry by focusing on specific use cases and offering unique

Threat of new entrants

The threat of substitute

Potter's five forces

The intensity of competition

The bargaining power of buyers

Bargaining power of suppliers

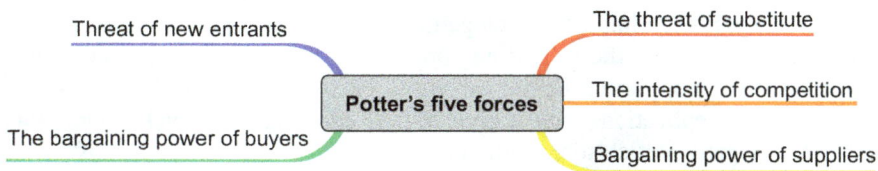

Figure 10: Porter's five forces (Source: Author's own work)

features or improvements over existing networks. For example, some networks have focused on scalability, privacy, or interoperability, attracting users and developers who value those features.

Bargaining Power of Suppliers: The suppliers in the blockchain industry are the developers and miners who create and validate transactions on the network. The bargaining power of these suppliers can be significant, especially in networks that rely on PoW consensus algorithms. However, the development of PoS consensus algorithms has reduced the power of miners, making the industry less dependent on a single group of suppliers. The development of Proof of Stake (PoS) consensus algorithms has reduced the power of miners, making the industry less dependent on a single group of suppliers. In PoS, validators are chosen based on their stake in the network rather than their computational power. This means that anyone with a sufficient stake in the network can participate in the validation process, reducing the concentration of power among a few large mining pools. Another factor reducing the bargaining power of suppliers in the blockchain industry is the open-source nature of the technology. The source code for many blockchain networks is publicly available, allowing anyone to contribute to the development of the network. This means that there is no single entity controlling the technology, reducing the power of any one group of suppliers.

The Bargaining Power of Buyers: This is an important factor to consider when analyzing the competitiveness of the blockchain industry. In the case of blockchain, buyers refer to users and investors who purchase tokens or use blockchain-based services. One of the main advantages of blockchain technology is its decentralized nature, which allows users to transact directly with each other without the need for intermediaries. This reduces the bargaining power of traditional intermediaries such as banks or financial institutions, as users can transact directly with each other. However, the bargaining power of buyers in the blockchain industry can still be significant, especially in networks that rely on token economics. In many blockchain networks, users must purchase tokens to use the network or access certain services. The demand for these tokens can be influenced by the number of users on the network, the availability of alternative networks or services, and the overall performance and security of the network.

Furthermore, buyers in the blockchain industry can also influence the direction of the technology by supporting or rejecting proposals for changes or upgrades to the network. This can impact the competitiveness of the network and the value of its tokens. In addition, the increasing popularity of decentralized finance (DeFi) has created new opportunities for buyers to exert influence in the blockchain industry. DeFi applications such as decentralized exchanges and lending platforms rely on user participation to provide liquidity and generate returns for investors. This has led to a proliferation of DeFi platforms, creating competition for users and investors.

However, the decentralized nature of blockchain technology also makes it difficult for buyers to exert significant bargaining power over the industry. The open-source nature of many blockchain networks allows developers to create new applications and services without the need for permission or approval from a central authority. This reduces the power of buyers to control the direction of the technology or restrict competition. Therefore, the bargaining power of buyers is an important factor to consider when analyzing the competitiveness of the blockchain industry. While the decentralized nature of the technology reduces the power of traditional intermediaries, buyers in the blockchain industry can still influence the demand for tokens and the direction of the technology.

The Threat of Substitute: Products are one of the five forces that determine the competitive landscape of an industry. It refers to the extent to which alternative products or services can satisfy the same customer needs as the original product or service. In the case of blockchain technology, the threat of substitute products is becoming increasingly relevant as new technologies emerge.

One potential substitute product for blockchain technology is traditional centralized databases. While centralized databases have been in use for decades, they lack the transparency and security of blockchain technology. However, some companies may still prefer using centralized databases due to their familiarity with the technology and lower costs. This creates a threat of substitution for blockchain technology. Another potential substitute product for blockchain technology is distributed ledger technology (DLT). DLT is like blockchain in that it is a decentralized ledger, but it has some key differences. For example, DLT can be permissioned, meaning only authorized parties can access the ledger. This makes DLT more suitable for enterprise use cases, where privacy and control are important. However, DLT lacks the level of decentralization and security that blockchain offers.

A third potential substitute product for blockchain technology is quantum computing. Quantum computing (Barmes et al., 2022) is a new type of computing that uses quantum mechanics to perform calculations. It has the potential to break the cryptography that underlies blockchain technology, making it less secure. However, quantum computing is still in its early stages of development, and it is unclear when it will become a viable substitute for blockchain technology.

Blockchain companies must focus on developing their technology and creating value for their customers to mitigate the threat of substitute products. They can do this by improving the security and scalability of their technology, reducing transaction costs, and exploring new use cases. By doing so, they can create a strong network effect that makes it difficult for substitute products to gain traction.

Therefore, the threat of substitute products is an important factor determining the competitive landscape of the blockchain industry. While there are potential substitutes for blockchain technology, such as traditional centralized databases, DLT, and quantum computing, blockchain companies can mitigate this threat by focusing on developing their technology and creating value for their customers. By doing so, they can maintain their competitive advantage and continue to drive innovation in the industry.

The Intensity of Competition: Rivalry is one of the five forces that determine the competitive landscape of an industry. It refers to the degree of competition between existing players in the market. In the case of blockchain technology, the intensity of competitive rivalry has been increasing as more companies enter the market and compete for customers and market share.

One factor contributing to the intensity of competitive rivalry in the blockchain industry is the low barriers to entry. Blockchain technology is open source and freely available, which means that anyone can develop their own blockchain and enter the market. This has led to a proliferation of blockchain projects and companies, each competing for a piece of the market.

Another factor contributing to the intensity of competitive rivalry in the blockchain industry is the lack of standardization. There are many different blockchain protocols and platforms, each with its own strengths and weaknesses. This makes it difficult for companies to build on each other's work and creates fragmentation in the market.

A third factor contributing to the intensity of competitive rivalry in the blockchain industry is the presence of dominant players. While there are many players in the market, a few companies have emerged as dominant players, such as Ethereum, Bitcoin, and Binance Smart Chain. These companies have a significant market share and influence over the direction of the industry.

Companies need to focus on differentiation and innovation to navigate the intensity of competitive rivalry in the blockchain industry. They can do this by developing unique features and capabilities that set them apart from competitors, such as faster transaction times, more secure smart contracts, or better user interfaces. They can also focus on building partnerships and collaborations with other players in the market, which can help to create network effects and increase market share.

Albeit the intensity of competitive rivalry is an important factor that determines the competitive landscape of the blockchain industry. While there are many players in the market and the barriers to entry are low, companies can navigate this intensity by focusing on differentiation, innovation, and collaboration. By doing so, they can gain market share and continue to drive innovation in the industry.

Therefore, Potter's Five Forces can be used to analyze the competitiveness of the blockchain industry. The high level of competition, the need for significant resources, and the bargaining power of buyers and suppliers are key factors in the industry. The unique features of blockchain technology, such as immutability, decentralization, and transparency, make it a compelling option for many use cases, but the industry still faces significant challenges in terms of adoption and scalability.

Using Design for Six Sigma for Blockchain Implementation

There are many quality management programs like Kaizen, Lean Management, Six Sigma, and DFSS. DFSS is a methodology used to design products or services that meet customer expectations and specifications (Jenab et al., 2018). DFSS uses various tools and techniques to identify problems and root causes, understand customer needs, and design solutions that address these issues.

Kaizen in Japanese means improvement, and it is a mindset of continuous improvement that can be applied to any process or activity (Lina and Ullah, 2019). Lean management is often used in manufacturing and production environments but can be applied to any business. It focuses on improving efficiency and quality by eliminating waste and reducing unnecessary steps in processes (Lawal et al., 2014). Six Sigma aims to reduce defects in products or services. It uses statistical methods to achieve quality improvement. DFSS is used to design products or services that meet customer needs and expectations while minimizing waste and maximizing value (Francisco et al., 2020). DFSS is often used in manufacturing, healthcare, and finance industries to optimize processes and products. In recent years, the use of DFSS has expanded to include implementing blockchain technology. Organizations can use DFSS to design blockchain implementations to ensure that the technology meets customer needs and delivers value (see Figure 11).

Blockchain technology has the potential to change various industries by providing secure, decentralized transactions. However, the implementation of Blockchain can be complex, and if not done correctly, can result in inefficiencies and security vulnerabilities. DFSS provides a structured approach to designing blockchain implementations that meet customer needs and deliver value. DFSS

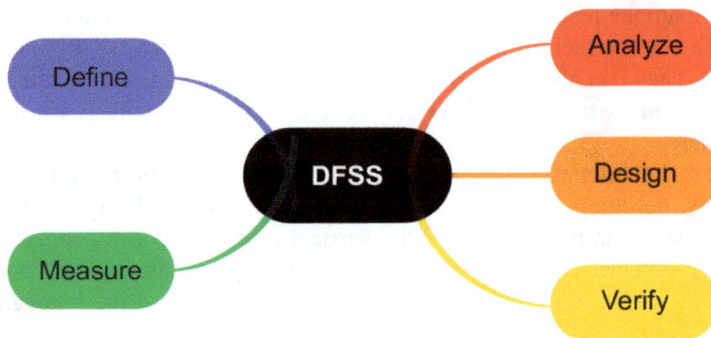

Figure 11: DFSS Model (Source: Author's own work)

is an extension of Six Sigma, a methodology to improve existing processes by identifying and eliminating defects. DFSS differs from Six Sigma in that it is used for designing new products, processes, or services, while Six Sigma is used for improving existing ones. DFSS uses a structured approach to design and development that is focused on meeting customer requirements and reducing variation in the final product or service (Chowdhury, 2002).

Define

In blockchain implementation, the first step in using DFSS is to define the customer requirements. Defining customer requirements is a critical step in the DFSS methodology because it ensures the project's goals and objectives align with the customer's needs. It involves gathering, analyzing, and prioritizing customer requirements, and developing a clear understanding of their expectations and needs. Customer requirements may include factors such as security, scalability, transparency, immutability, and ease of use.

To define customer requirements for blockchain implementation, it is essential to engage with stakeholders, including customers, partners, and regulators. This engagement can be achieved through surveys, focus groups, and interviews. By engaging with stakeholders, businesses can gain insights into what their customers expect from blockchain implementation and the benefits they hope to gain from it. Once the customer requirements have been identified, the next step is to prioritize them based on their impact on the project's success. Prioritization involves ranking requirements in order of importance and identifying those that are critical to the project's success. For example, security may be a critical requirement for blockchain implementation in the financial industry, while ease of use may be more critical in the healthcare industry.

After prioritization, businesses can then develop a design concept that meets the customer's requirements. The design concept involves creating a detailed plan for implementing blockchain technology that addresses the prioritized customer requirements. This plan may include the technology architecture, software and hardware requirements, and integration with existing systems.

In conclusion, the first step in using DFSS for blockchain implementation is to define the customer requirements. Defining customer requirements is a critical step that ensures the project's goals and objectives are aligned with the customer's needs. By engaging with stakeholders, prioritizing requirements, and developing a design concept that meets customer requirements, businesses can ensure successful blockchain implementation that meets the customer's expectations and needs.

Measure

In the Measure phase of DFSS, data is collected and analyzed to determine the current state of the process or product. In blockchain implementation, the Measure phase is critical as it helps to identify the critical-to-quality (CTQ) characteristics that are essential to meeting customer requirements. The CTQ characteristics of a blockchain solution may include factors such as security, scalability, transaction speed, network throughput, and consensus mechanism. To collect the required data, it is essential to define the scope of the project and identify the relevant stakeholders. The stakeholders may include customers, end-users, developers, regulators, and other parties involved in the blockchain ecosystem. The data collection may involve conducting surveys, focus groups, or interviews with stakeholders to gather their opinions and feedback. Additionally, data from existing blockchain networks may be analyzed to identify trends, patterns, and performance metrics.

The analysis of the collected data involves using statistical tools and techniques to identify the sources of variation and defects. In blockchain implementation, the sources of variation may include issues such as network congestion, security vulnerabilities, or interoperability challenges. The analysis may involve using tools such as Pareto charts, root cause analysis, and failure mode and effect analysis (FMEA) to identify the critical factors that affect the quality of the blockchain solution.

The data collected and analyzed in the Measure phase provides the foundation for the subsequent phases of DFSS, such as Analyze, Design, and Verify. The analysis of the data helps to identify the customer requirements and CTQ characteristics, which are essential in designing a blockchain solution that meets the needs of the stakeholders. Additionally, the data analysis provides insights into the potential risks and challenges that may affect the quality of the blockchain solution, allowing the project team to develop mitigation strategies to address them.

Analyze

The Analyze phase of DFSS involves identifying the root causes of the problems identified in the Measure phase. In the context of blockchain implementation, the Analyze phase aims to identify the factors that impact the performance, security, and scalability of the blockchain solution. This is achieved through the use of statistical analysis tools and techniques such as hypothesis testing, regression analysis, and design of experiments (DOE).

One of the critical factors affecting the performance of a blockchain solution is the consensus mechanism used. The consensus mechanism determines how transactions are validated and added to the blockchain. Different consensus mechanisms have their strengths and weaknesses, and the choice of the consensus mechanism depends on the specific requirements of the application. In the Analyze phase, the project team can use DOE to evaluate the performance of different consensus mechanisms and identify the optimal solution.

Another critical factor affecting the security of a blockchain solution is the vulnerability of smart contracts. Smart contracts are self-executing programs that are executed on the blockchain. However, if these smart contracts contain bugs or are not designed correctly, they can be exploited, leading to significant financial losses. In the Analyze phase, the project team can use hypothesis testing to identify the critical factors contributing to the vulnerabilities in smart contracts and develop strategies to mitigate them.

The scalability of a blockchain solution is another critical factor that can be addressed in the Analyze phase. Scalability refers to the ability of the blockchain network to handle an increasing number of transactions without compromising its performance. One of the challenges of scaling blockchain solutions is the increasing size of the blockchain ledger. In the Analyze phase, the project team can use regression analysis to identify the factors contributing to the size of the blockchain ledger and develop strategies to manage its growth.

In conclusion, the Analyze phase of DFSS plays a critical role in ensuring the successful implementation of blockchain solutions. Through the use of statistical analysis tools and techniques, the project team can identify the root causes of the problems identified in the Measure phase and develop strategies to mitigate them. By using a structured approach such as DFSS, businesses can develop and deliver high-quality blockchain solutions that meet customer requirements and provide value to the stakeholders.

Design

The Design phase of DFSS involves creating a detailed plan for developing and implementing the solution. In the context of blockchain implementation, the Design phase aims to develop a blockchain solution that meets the customer requirements identified in the Measure phase and addresses the root causes of the problems identified in the Analyze phase. This is achieved through the use of tools and techniques such as process mapping, failure mode and effect analysis (FMEA), and design of experiments (DOE).

One of the critical factors to consider in the Design phase of blockchain implementation is the choice of the blockchain platform. Different blockchain platforms have their strengths and weaknesses, and the choice of the platform depends on the specific requirements of the application. For example, if the application requires high-speed transaction processing, a platform such as EOS or Ripple may be suitable. If the application requires a high level of security, a platform such as Ethereum may be more appropriate.

Another critical factor to consider in the Design phase is the design of smart contracts. Smart contracts are self-executing programs that are executed on the blockchain. In the Design phase, the project team can use FMEA to identify the potential failure modes of smart contracts and develop strategies to mitigate them. Additionally, the project team can use DOE to optimize the performance of smart contracts and ensure that they meet customer requirements. The scalability of the blockchain solution is another critical factor that can be addressed in the Design phase. In the Design phase, the project team can use process mapping to identify the key processes involved in scaling the blockchain solution and develop strategies to optimize them. Additionally, the project team can use DOE to identify the critical factors that contribute to the scalability of the blockchain solution and develop strategies to improve it.

In conclusion, the Design phase of DFSS plays a critical role in ensuring the successful implementation of blockchain solutions. By using tools and techniques such as process mapping, FMEA, and DOE, the project team can develop a blockchain solution that meets the customer requirements and addresses the root causes of the problems identified in the Analyze phase. By using a structured approach such as DFSS, businesses can develop and deliver high-quality blockchain solutions that provide value to the stakeholders.

Verify

The Verify phase of Design for Six Sigma (DFSS) involves validating the solution developed in the Design phase and ensuring that it meets the customer's requirements. In the context of blockchain implementation, the Verify phase aims to ensure that the blockchain solution developed meets the performance, security, and scalability requirements identified in the previous phases. This is achieved using statistical analysis tools and techniques such as hypothesis testing, regression analysis, and control charts.

One of the critical factors to consider in the Verify phase of blockchain implementation is the performance of the blockchain solution. In the Verify phase, the project team can use hypothesis testing and regression analysis to validate the performance of the blockchain solution and ensure that it meets the performance requirements identified in the Design phase. For example, the project team can validate the speed of transaction processing and the latency of the blockchain network.

Another critical factor to consider in the Verify phase is the security of the blockchain solution. In the Verify phase, the project team can use statistical analysis tools and techniques to validate the security of the blockchain solution and ensure that it meets the security requirements identified in the previous phases. For example, the project team can validate the robustness of the smart contracts and the encryption algorithms used in the blockchain network. The scalability of the blockchain solution is another critical factor that can be addressed in the Verify phase. In the Verify phase, the project team can use control charts to monitor the performance of the blockchain solution and ensure that it can handle an increasing number of transactions without compromising its performance. Additionally, the project team can use statistical analysis tools and techniques to identify any bottlenecks or limitations in the scalability of the blockchain solution and develop strategies to mitigate them.

In conclusion, the Verify phase of DFSS plays a critical role in ensuring the successful implementation of blockchain solutions. By using statistical analysis tools and techniques such as hypothesis testing, regression analysis, and control charts, the project team can validate the performance, security, and scalability of the blockchain solution and ensure that it meets customer requirements. By using a structured approach such as DFSS, businesses can develop and deliver high-quality blockchain solutions that provide value to the stakeholders.

Conclusion

Blockchain technology can improve transparency and traceability in the supply chain by providing a decentralized, tamper-proof ledger of transactions that all participants can access and verify. This allows for greater visibility into the movement of goods and materials from their origin to their final destination.

By creating an open architecture for sharing information, blockchain technology can also enable more efficient and secure communication between different parties in the supply chain, such as manufacturers, distributors, and retailers. This can help to reduce costs, increase speed, and improve overall supply chain performance.

Moreover, blockchain technology can enable the creation of smart contracts, which are self-executing contracts that automate the terms of an agreement between parties. This can reduce the need for intermediaries and paperwork and improve the efficiency and accuracy of transactions.

Overall, the transparency, traceability, and efficiency benefits of blockchain technology can be particularly useful in industries where supply chain management is critical, such as in food and pharmaceuticals, where safety and regulatory compliance are top priorities.

The scope of blockchain technology goes beyond just the control and sharing of data with stakeholders on the blockchain. Blockchain technology is a decentralized digital ledger that stores transactions in a secure and tamper-proof manner. It operates on a peer-to-peer network, where all participants have equal rights and responsibilities. The technology provides a way to establish trust among parties without the need for intermediaries, such as banks or other financial institutions.

One of the primary benefits of blockchain technology is its ability to provide a secure and transparent way to manage and transfer data. This includes data related to transactions, identity, property ownership, and more. By using cryptography and consensus mechanisms, blockchain technology ensures that data stored on the blockchain cannot be altered or deleted without the agreement of the network.

Beyond data control and sharing, blockchain technology can also enable the automation of complex business processes through smart contracts. Smart contracts are self-executing digital agreements that can be programmed to trigger actions based on predefined conditions. This can help to reduce the need for intermediaries and streamline processes. In addition, blockchain technology has the potential to revolutionize a variety of industries, including finance, supply chain management, healthcare, and more. It can provide new ways to verify and authenticate transactions, reduce fraud and errors, and increase transparency and accountability.

Overall, the scope of blockchain technology is vast and continues to evolve as new use cases and applications are developed.

Logical statements, also known as conditional statements or if-then statements, can be embedded in blockchain technology to automate transactions and create efficiencies in various processes.

By using smart contracts, which are self-executing contracts with the terms of the agreement written into code, logical statements can be used to automate processes that previously required human intervention. For example, if a certain condition is met (such as the receipt of a payment), then the smart contract can automatically trigger another action (such as the release of a product or service). This automation can reduce operating costs by eliminating the need for intermediaries or manual processes, as well as increase the efficiency of transactions by reducing the time required for verification and approval. Furthermore, blockchain technology provides transparency and security through its decentralized and immutable nature, ensuring that transactions are recorded accurately and cannot be tampered with. This can further reduce the risk of errors and fraud, adding to the overall efficiency of the blockchain system.

Blockchain technology can simplify various aspects of human life, including those in the sectors you mentioned such as energy, agriculture, health, construction, manufacturing, and supply chain. Blockchain technology is a distributed ledger technology that allows for the secure and transparent recording of transactions without the need for intermediaries. This means that information can be shared in a decentralized and trustless manner, reducing the need for complex and expensive systems and intermediaries. In the energy sector, blockchain technology can be used to enable peer-to-peer energy trading and management, while in agriculture, it can be used to track the origin and quality of products. In healthcare, blockchain technology can be used to securely store and share patient data, while in construction, it can be used to ensure the integrity and safety of building materials and structures.

In manufacturing, blockchain technology can be used to improve supply chain transparency and traceability, while in the supply chain, it can be used to track goods and ensure their authenticity. Overall, blockchain technology has the potential to simplify and improve many aspects of our lives, making processes more efficient, secure, and transparent.

Chapter Summary

Blockchain technology has several characteristics that make it unique and useful for various applications. Here is a brief explanation of the characteristics:

- **Traceability:** Every transaction in a blockchain network is recorded in a permanent and unchangeable way. This allows for easy tracking and auditing of assets, information, or any other data that is stored on the blockchain.
- **Transparency:** The data stored on a blockchain is visible to all network participants, and any changes or updates made to it can be seen and verified by anyone. This creates a high level of transparency and accountability in the network.
- **Time-stamped:** Each transaction or block added to the blockchain is time-

stamped, meaning that the exact date and time of when it occurred is recorded. This helps to create a clear and accurate record of events and transactions.

- **Secured:** The data stored on a blockchain is secured through advanced cryptographic techniques, making it virtually impossible to alter or tamper with the data once it has been added to the network.
- **Consensus-driven:** In a blockchain network, all participants must agree on the validity of each transaction before it is added to the blockchain. This consensus-driven approach helps to ensure the accuracy and integrity of the network.
- **Censorship-resistant:** Because blockchain networks are decentralized, it is difficult for any one entity to censor or manipulate the data stored on the network. This makes it a powerful tool for protecting freedom of speech and preventing censorship.

Blockchain technology has the capability to offer several value drivers, including transparency and reduced administrative costs, lowered risk of fraud, and improved control of subcontracted manufacturing.

- Transparency is one of the key benefits of blockchain technology, as it enables all parties involved in a transaction to have access to a tamper-proof and permanent record of the transaction. This can help to increase trust and reduce the risk of fraud, as it is much harder to manipulate or falsify records on a blockchain.
- Reduced administrative costs are another potential value driver of blockchain technology, as it can help to streamline processes and eliminate the need for intermediaries. By using blockchain technology, transactions can be processed faster and more efficiently, which can result in significant cost savings.
- Lowered risk of fraud is also a major benefit of blockchain technology. The decentralized nature of blockchain means that transactions are validated and recorded by a network of participants, making it much harder for any single party to manipulate the transaction or engage in fraudulent activity.

Blockchain technology can also improve control of subcontracted manufacturing by providing a tamper-proof and permanent record of all transactions and interactions in the supply chain. This can help to increase visibility and transparency, reduce the risk of fraud, and improve overall supply chain management.

- **Transparency:** Blockchain technology enables transparency by creating an immutable and publicly accessible ledger of transactions. This means that all parties involved in a transaction can view and verify the details of the transaction, which can increase trust and reduce the potential for fraud or disputes.

- **Credibility:** Blockchain technology can increase credibility by providing a secure and tamper-proof record of transactions. This can help to verify the authenticity of assets, such as digital identities, intellectual property, or financial instruments.
- **Risk mitigation:** Blockchain technology can mitigate risk by enabling smart contracts and other automated processes that reduce the need for intermediaries and manual processing. This can help to reduce errors, delays, and other risks associated with traditional methods of conducting business.
- **Stakeholder engagement:** Blockchain technology can facilitate stakeholder engagement by providing a platform for decentralized decision-making and collaboration. This can enable more inclusive and democratic governance models, as well as greater participation and accountability among stakeholders.

Blockchain technology can potentially provide a range of intangible benefits that can help to improve trust, reduce risk, and increase engagement in a variety of industries and contexts.

Take Away Lessons from the Chapter

1. Blockchain technology provides improved transparency and traceability of material throughout the supply chain and creates an open architecture for sharing information in near real time for all users.
2. Supply chain visibility for organizations can benefit from enhanced visibility throughout the supply chain system. Some important benefits include reduction in administrative costs and identifying counterfeit products.
3. Blockchain utilizes the use of logical statements; thus, embedding logical statements in a blockchain infrastructure can automate transactions and link effectively with smart contracts to create operational efficiencies and reducing operating costs.
4. Blockchain technology promotes transparency, crediibility, risk mitigation , and stakeholder engagement, these are some sound and tangible benefits of blockchain technology.
5. Blockchain's decentralized premises create a secure, immutable, and verifiable network for peer-to-peer transactions that are quick, safe, and transparent.
6. DFSS provides a structured approach to designing blockchain implementations that meet customer needs and deliver value.
7. DFSS is an extension of Six Sigma, a methodology used to improve existing processes by identifying and eliminating defects. DFSS differs from Six Sigma in that it is used for designing new products, processes, or services, while Six Sigma is used for improving existing ones.

Chapter Exercises

1. Develop a SWOT analysis for an organization of your choosing. What are some of the strengths, weaknesses, opportunities, and threats that you have identified? Why are they important?
2. Why are methodologies such as Six Sigma and DFSS important when considering new technology? How do you think blockchain can help integrate preexisting proven business methodologies?
3. What are some of the benefits of integrating blockchain technology? What are some of the disadvantages of integrating blockchain technology and what should organizations consider before implementation?

Blockchain and Beyond

There is an exorbitant amount of data that exists due to data sharing. Currently, big data is operated through IoT, and IoT infrastructures and can be leveraged to improve life by creating utility and efficiency. Although IoT infrastructures and devices generate fine-grained data that can be analyzed to pointedly promote the development and optimization of technology, Isolatable resources such as electricity, mobile data, and large cloud-based stored data are often used to compute efficiencies. Optimization can be achieved through shared datasets to develop from reactive processes to proactive ones. The use of Blockchain technology platforms addresses privacy concerns and creates utility. In the data sanitizing process, blockchain application facilitates a framework that creates checks and balances while creating visibility of shared data to authorized users. On a global scale, digital transformation is the main key to the development of all sectors of the economy. Blockchain technology is a disruptive innovation that has changed the perception of many sectors. Compared to traditional data storage, blockchain offers an innovative solution to drive the optimization and promotion of transactions in all sectors of industries. The rapid evolution of global business architecture creates various opportunities and challenges. As global industries continue to grow, competition and the race for resources continue to demand robust solutions for complex business needs. Innovative solutions that create value with data and information are needed in this era of big data. Blockchain technology is of great significance to promote industrial transformation and technological innovation.

Introduction

There are huge amounts of data that is generated from devices such as mobile applications, GPS navigation systems, urban traffic cameras, crime detections, hospitals, and utilities. All this data is operated through IoT, and IoT infrastructures can be leveraged to improve life by creating utility and efficiency. IoT infrastructures and devices generate fine-grained data that can be analyzed to pointedly promote the development and optimization of technology. Divisible resources such as electricity, mobile data, and large cloud-based stored data are often used to compute efficiencies. To achieve optimization, the shared datasets are used to develop from reactive processes to proactive ones. Analyzing fine-grained data such as vehicle trajectory data, resource allocation data, and surveillance camera data would drastically benefit consumers and organizations. However, there are severe privacy concerns and security concerns surrounding the process of sharing and collecting information to achieve utility for the users. If sensitive data is directly released or shared these datasets would pose severe privacy problems for the consumers (Xuegang et al., 2011; Rizzo et al., 2015). Data tracking and pattern development, although not illegal, create privacy concerns for the consumer. To address privacy concerns and to create efficiencies in data management and distribution, a technology that creates specific types of security parameters is required to minimize privacy risks. Presently some privacy-enhancing technologies (PET) are used to sanitize data; however, there are major limitations to these technologies (Balon et al., 2022). The present technology either lacks formal privacy notions to quantify and bound privacy risks or has very limited utility. The limitations can be related to data sharing where only a small sequence of locations or aggregated information is released creating restricted visibility and utility.

Blockchain technology platforms address privacy concerns and create utility. In the data sanitizing process, blockchain application facilitates a framework that creates checks and balances while creating visibility of shared data to authorized users. During the data exchange process, some buyers and sellers may not trust each other without a market mediator. Fortunately, blockchain technology and the use of smart contracts can ensure digital agreements among mutually distrustful agents. Blockchain consensus protocol facilitates smart contract execution to achieve strong integrity. A hybrid blockchain platform ensures privacy, honesty, and high efficiency among users. Among the many utilities of blockchain, blockchain-based platforms ensure the integrity and availability of the transaction that helps data recovery if the execution protocol is interrupted. In its first, simplest form smart contracts were built into cryptocurrencies that are traded on a decentralized network of peers, and the surrounding all transactions are shared via a public ledger and further fortified through a consensus protocol on the blockchain.

A blockchain maintains a distributed ledger via consensus protocol and generates validity. A popular on-chain design is Hyperledger Fabric (Maheswaran and Basar, 2003). A Hyperledger Fabric is designed to distribute ledger technology with multiple modules for blockchain platforms. Blockchain as a distributed ledger platform includes a highly modular and configurable architecture. Supply chain management (SCM) is a systematic approach to managing the flow of goods, services, information, and money that requires collaboration across many business units and participants (Sutia et al., 2020). SCM is a dynamic and complex process that is integral to the overall success of an organization (Gloet and Samson, 2018). Innovations in technology have encouraged SCM applications to create strategic inter-organizational collaborations (Min et al., 2019). The supply chain's main function is to contribute to organizational efficiency and effectiveness (Breite and Koskinen, 2014; Bravo et al., 2016). The overall performance of organizations is directly related to the effectiveness of SCM functions like inventory management, purchasing, and logistics. The supply chain process is heavily laden with manual work and human decision-making.

Tourism and Fake News

On a global scale, digital transformation is the main key to the development of all sectors of the economy. Blockchain technology is a disruptive innovation that has changed the perception of many sectors. Compared to traditional data storage, blockchain offers an innovative solution to drive the optimization and promotion of transactions in all sectors of industries. The digital era has been gaining momentum from the adoption of massive online transactions by consumers. The convenience of digital technology has catapulted digital services and radically changed economies globally. Digital services have quickly become the new sensation and different industrial applications in education, sciences, supply chain, logistics, finance, and law enforcement have made an enthusiastic effort to seek out blockchain technology to repurpose digital data.

The global tourism industry often must accommodate changing, now popular global trends. Blockchain technology opens opportunities for Global tourism that remain misunderstood. The global market for the distribution of tourist services is extremely fragmented and is currently facilitated through many intermediaries (Panina et al., 2022). For instance, a simple hotel booking would have two main channels for booking: a channel manager – that caters to small and large travel agencies and direct to consumer websites. However, there can be more intermediaries involved on a smaller scale that deal directly with the larger online travel agencies creating more layers. Global Distribution System (GDS) is the largest booking system in the world. Service providers are permitted to upload information on room availability, airline tickets, car rentals, etc. on their databases. Therefore, when Delta Airlines, an active participant on the GDS network conducts business using GDS systems, the cost of the ticket is increased due to subscription fees to the GDS platform. Similarly, Online Travel Agencies (OTA) is the largest online travel agency that includes booking sites such as Booking.com, Expedia. com, etc. OTAs can be simultaneously connected to GDS to access air tickets, rooms, car rentals, etc., and can earn 10–30% commission, which is also an additional expense that increases the cost of tourist services. Channel Managers can connect to both OTA and GDS systems and allow hotels to be participants in managing sales channels and bookings from one central location and assume there is a commission included in the price of the rooms, which is insignificant compared to GDS and OTA commissions. It is painfully clear that the global tourism industry operates in an extremely fragmented and complex intermediary market that significantly increases the price of tourism to the direct consumer.

Decentralizing global tourism activities and reforming the use of intermediaries can create massive value for the industry. Valuable data shared among the participants provides a key strategic direction to revolutionize the role of the field of travel and offers tremendous potential as a new digital mechanism to organize the shared data. A blockchain based solution will provide a platform for booking travel services on an open global distribution system that is controlled by the participants on the network and eliminates the power of single ownership. Blockchain enables a decentralized approach to databases, storing information, and connecting to centralized servers to present a repository of listed and ordered records or blocks. Blockchain technology has emerged as the front-runner in decentralized technology (Khan and Masood, 2022).

The access to data provides service providers with a technical opportunity to upload transactional data to the system conveniently and cost-effectively. Not only is the safety of the data maintained by creating a decentralized system, substantial commissions and extra charges can be avoided from intermediaries. In addition, blockchain significantly simplifies the identification of passengers while maintaining personal data securely and improves the mechanism for tracking luggage, and facilitates mutual settlements between the consumer and travel agencies, airlines, etc. Several companies in the travel industry that have accepted blockchain technology to differentiate themselves. WindingTree is a decentralized distribution network for tourist services that improves the quality of consumer communication with the counterparty using smart contracts and removes unnecessary intermediaries thereby reducing costs and simplifying the transaction process. Similarly, Concierge is a mobile application that allows users to book hotels without paying commissions and pay for their stay using cryptocurrency. The willingness of industries to apply blockchain technology in their business practices proves that the process of digitization of economies has already leapfrogged to implement innovative technology. There is a clear alteration in consumer behavior that must be met by industries.

The rapid evolution of global business architecture creates various opportunities and challenges. As global industries continue to grow, competition and the race for resources continue to demand robust solutions for complex business needs. Innovative solutions that create value with data and information are needed in this era of big data. Blockchain technology is of great significance to promote industrial transformation and technological innovation (Kumar et al., 2022; Wang et al., 2018; Xue, 2022). The world's scientific and economic infrastructure is constantly being reshaped. New-generation technological theories such as blockchain, artificial intelligence, and big data are not just limited to information technology. The progressive maturation of the new generation of technology is bringing vitality and removing traditional barriers by transforming and upgrading industries and existing models (Khan et al., 2020; Zhang, et al., 2020). Blockchain technology's most attractive characteristics are immutability, uniqueness, decentralization, and application of smart contracts (Ali, et al., 2020; Jia et al., 2019). The global flow of technology, money, and information has broken barriers in the global economy and supports the development of globalization.

The prevalence of globalization creates more difficult circumstances for organizations to manage their data (Chang et al., 2020; Du et al., 2019). The differences in local to central policies create more confusion in business management for organizations. To pursue an efficient management model, many organizations have adopted re-engineered models to establish an independent individual agency to engage in services. There have been many implementations in various industries that have been supported by blockchain-based structures to reconstruct old processes, engineer new processes, and share data in a secure, high quality, and low-risk sharing manner (Cainelli, 2012; Luecke and Schneiderheinze, 2017). Using the full potential characteristics of blockchain technology breaks the distrust among individuals and achieves business automation in a secure environment. The decentralization mechanism facilitates more trust and business transactions are conducted in a trusted, traceable, and validated environment with data integrity that can be shared with all users (Sun et al., 2016; Underwood, 2016). Overall integrating financial transactions on a blockchain-based database promotes customer growth rate, financial service satisfaction, and risk control level. In addition, financial organizations can generate business process optimization, reduce business error rates, and increase industrial business processing efficiencies. Access to one's financial information has become more critically important and in the current arena of finance, consumers must have access and the ability to direct their data. The information technology era that has survived the credit crisis needs to remove vulnerabilities from the current architecture. Blockchain technology relies on a distributed encrypted ledger methodology and can have a significant impact on the current finance industry ecology and operating model.

Food safety concerns are on the rise. Consumers are actively demanding food sourcing transparency to build trust and credibility in the consumption of their food. There have been many challenges faced in the environment of the marine market of products that include marine pollution, wastewater discharge, uncertainty, untrustworthy sourcing, and overfishing (Halik et al., 2018; Harris et al., 2021; Lestari and Trihadiningrum, 2019; McIlgorm et al., 2022).

The government plays an important role in representing public benefit by controlling environmental pollution. Existing government regulations are not successful in regulating environmental problems effectively. However, the development of regulation and compliance is a key responsibility of the regulators to control pollution and construct governance systems for punishing offenders (Kosajan et al., 2018; Xu et al., 2021). Therefore, government regulators must consider adopting technology that improves the accuracy of regulatory controls and helps to regulate policies for the benefit of the consumer and protect the earth (Venkatesh et al., 2020). Innovative technologies that reduce the burden of ineffective manual and problematic transaction information must be explored. The incorporation of blockchain technology in the food safety arena has been very successful. According to Howson (2020), blockchain has incredibly transformed and improved the level of marine conversation and the global fishery supply chain by incentivizing waste collection and mitigating ocean pollution. Parts of global logistics for products, especially, food products are dependent upon cumbersome manual paperwork part of customs clearing. Regulators can implement blockchain technology to realize market governance and improve accountability in the maritime supply chain including punishment management (Banach, 2020; Liu et al., 2021; Wang et al., 2021). The role of blockchain technology can be galvanized to improve social and environmental benefits. The inherent features of blockchain technology can improve the trust level of products and provide sourcing information for environmental protection. Information asymmetry and lack of transparency have hindered food safety regulations, blockchain provides a viable solution for both transparency and tracing of the product that creates better regulations and food safety.

The importance of food regulations cannot be undermined. Organizations must interact with government agencies as it is a significant and reasonable part of doing business. Blockchain technology creates a strategic advantage for organizations that seek to incur efficiencies in their operations and thoughtfully include regulators to oversee entitled information. To understand the impact of social, environmental, and food safety factors let's illustrate a scenario. Imagine the polluted product that has been brought to the market for sale. Two scenarios can occur.

1. A blockchain-based organization will absolutely refuse to purchase the polluted product. This is because the transparency of the origin and therefore pollution are all records that are readily available to the organization to make its decision.
2. An organization that does not have blockchain technology purchases the polluted product and then sells the product for a lower price unless the government intervenes.

Advanced information technologies such as artificial intelligence (AI) and blockchains can ensure efficacy, regulate information asymmetry, and increase trust. Consumption of mass media information has become a challenge due to fake news. Fake news is the deliberate spreading of false information disguised as being authentic (Lacity, 2021). The most common fake news is fabricated by political, anti-scientific, and economic communities. Politicians aim to spread false information about other politicians. Anti-scientific news such as aliens, vaccines, diseases, etc. and economic fake news aims to spread false information about stock prices, market manipulations, etc. (Lazer et al., 2018). Although fake news is not a phenomenon, it is the cause of distrust. The 21st century, also the age of digital technologies has amplified the creation and distribution of fake news. Today, social media platforms are a huge distribution channel for fake news. Fake news is a serious societal problem as it undermines the public's confidence. There is an overload of information that is available, and it is extremely hard to differentiate between what is fake and what is true. To combat fake news, social media platforms rely on AI to provide solutions. Many AI solutions detect and flag fake news through algorithms and word checking. In addition to AI tracking tools, the industry is actively seeking blockchain-enabled solutions to provide services such as establishing content authenticity, tracking, blacklisting imposters, and spotting deepfakes (Harrison and Leopold, 2021).

The blockchain-based approach has tremendous potential in battling fake news and digital disinformation. Blockchain utilizes a decentralized, immutable ledger to record information that is constantly confirmed and reconfirmed by all users, making it almost impossible to alter information after it has been created. Blockchain's ability to provide decentralized validation with a clear chain of custody makes it an important solution in all sorts of forms and industries. One reason why deepfakes are hard to combat is that disinformation does not follow consistent standards or best practices to identify and track manipulated media across digital platforms. Blockchain provides greater transparency into the lifecycle of the content and can offer a mechanism to restore trust in our digital ecosystem. There are three main techniques by which blockchain technology can address the challenges of digital information: Verification of origin, secure maintenance of online identity and reputation, and incentivizing high-quality content.

Blockchain technology can be used to combat digital misinformation by tracking and verifying sources for different online media platforms. Publishers can use blockchain technology to create and register their content, providing greater context and transparency into how the data was created. Open-source content generally is more prone to misuse. Therefore, open-source academic and research organizations can create non-malicious applications such as educational videos and track and verify individuals who access the blockchain.

Traditionally publishers have been the primary source of reputation behind the content. However, in mainstream digital overload, even reputable publications are increasingly incentivizing engagement over clarity. Readers are likely to get information from social media which seriously impedes their ability to distinguish between credible journalist outlets and interest-driven propaganda. A blockchain-based platform can verify the identity of content creators and trace their reputation for accuracy, eliminating the need for a centralized institution.

The model for incentivizing high-quality content needs to be flipped. One of the most challenging aspects of promoting accurate information is the strong incentivization to drive clicks at all costs to sensationalize the content. Blockchain's use of smart contracts offers a built-in mechanism to automate payment for verified content with predefined quality standards. This is a very powerful tool to combat the current model of incentivizing low-quality content.

Global Supply Chain Adoption

An organization may be part of multiple blockchains. Therefore, the industry must understand how these blockchains interact with each other. While there are several barriers to blockchain adoption for globally sustainable chain management, there is evidence of successful integration of blockchain technology in the supply chain. Kamble et al. (2020) reported that regardless of impediments, blockchain applications in agricultural supply chains have yielded positive results through the reduction in transaction costs, information sharing, and data security. Additionally, participants who choose to share data on a blockchain must agree on standardized processes, formats, and levels of detail regarding the data sharing among peers. Establishing clear governance rules for blockchain platforms creates trust at the baseline to work efficiently, a blockchain must have an appropriate governance structure and clear rules for decision-making and conflict management.

Theoretical Baseline

According to the theory of diffusion of innovation (DOI), the adoption of new technology is based upon five criteria: the complexity of the innovation, compatibility with the organization, benefits, and disadvantages as compared to the legacy system, observability, and trialability (Rogers, 2010). Gunduz and Das (2020) posit that there are five groups of enablers that include three characteristics of innovation: relative advantage, compatibility, and flexibility. Furthermore, organizations and external factors all play an important role in shaping the decision behind the adoption of blockchain technology in the supply chain.

Advantages of Blockchain in Supply Chain

Innovative technology such as blockchain creates economic benefits as compared to the preexisting legacy technology that it replaces. Kapoor et al. (2014) reported that the advantages of technology adoption are often positively correlated. Literature regarding blockchain technology unequivocally supports the premise that blockchain technology ensures better data integrity and safety and can improve data availability for multiple users in a supply chain setting. There is also consensus in the market regarding the use of this technology to reduce the cost of transactions between members. Sharing of records in supply chain management is a unique advantage of blockchain technology. Data sharing among blockchain participants in the supply chain provides increased transparency, and visibility of information through the entire supply chain thereby creating efficacies in the internal operational processes by creating internal traceability and details of operational data. In their studies, Kapoor et al. (2014) discovered that lower transaction cost, customer interest in traceability, and product-related data on blockchain were the two top prominent enablers of blockchain technology adoption in supply chain management. Johnson's (2021) rhetoric validated that organizations' commitment and cooperation among supply chain participants to share data, confidentiality, and system governance are all important facets for adopting blockchain technology in the supply chain. The research also validated that there is a push from consumers to understand product-related data, thus this is the primary source of pressure for organizations to adopt blockchain technology to meet consumer demands.

The adoption of blockchain technology in supply chain management has several economic benefits compared to legacy technology. The use of blockchain technology ensures better data integrity and safety, improves data availability for multiple users, and reduces the cost of transactions between members. Additionally, blockchain technology enables data sharing among supply chain participants, which provides increased transparency and visibility of information throughout the entire supply chain, thereby creating efficacies in internal operational processes. Lower transaction costs and customer interest in traceability and product-related data on blockchain were found to be the top prominent enablers for blockchain technology adoption in supply chain management. Commitment and cooperation among supply chain participants to share data, confidentiality, and system governance were also identified as important facets of adopting blockchain technology in the supply chain. Finally, there is a push from consumers to understand product-related data, which is a primary source of pressure for organizations to adopt blockchain technology to meet consumer demands.

According to the IBM Blockchain Transparent Supply Solution Brief (2022a), there is limited visibility of data regarding the movement of goods from suppliers to customers, and the current data that is available is mostly recorded in paper-based transactions. Although blockchains are complex, they could transform the supply chain by providing a trusted and controlled avenue of data sharing in real-time visibility. Industry leaders and visionary companies are adopting blockchain to satisfy their most challenging problems. Solutions like IBM architecture for building blockchain in supply chain management offer a platform that enables specific transparency across multiple supply chain partners and provides analytics, location, product movement, and condition that is built out per the organization's customized determination. IBM Blockchain Transparent Supply Solution Brief (2022a) discovered that while 78% of companies using blockchain technology increase visibility as the main benefit of the technology, 37% of companies surveyed have considered blockchain technology as a solution.

In the supply chain, blockchain-based technology provides collaboration, data sharing, and visibility to all users across the supply chain ecosystem. The blockchain functions by vetting approved participants to connect and share data on a distributed ledger to provide transparency and truth about goods as they make their way through the supply chain. Participants add documentation such as certifications and any other relevant documentation to the ledger. Once the data is added to the ledger it cannot be manipulated. These are some strong properties of blockchain that make it a desirable technology for trusted information sharing of data across globally participating firms and help in the prevention of fraud. As organizations choose, the organizations are in control of whom to share data with. Data stored on the blockchain is protected with the highest level of tamper-resistant encryption. Thus, allowing participants to access data equitably and improving efficacy, building trust, and removing friction. The robust mechanism of blockchain technology allows users the speed to handle millions of transactions proficiently. Blockchain allows organizations to track the physical flow of goods across companies with high speed and insight, combine data that is typically siloed to leverage new business and automate processes across companies to reach elevated supply chain efficiency. Some noticeable advantages of transparent supply are enabling end-to-end visibility to empower all participants in the supply chain and to create new strategic business value. Organizations can drive buyer engagement to increase their market share by engaging and empowering buyers to build trust in their brand by showing the consumer where the product was made and demonstrating support for ethical and sustainable production. Blockchain technology facilitates transformation and business model optimization by leveraging supply chain visibility and improving predictions and forecasting through real-time data to optimize inventory. In addition, smart contracts automate responsiveness through the supply chain.

Blockchain technology has the potential to revolutionize the supply chain industry by enabling greater transparency, efficiency, and trust. By leveraging a distributed ledger system, all participants in the supply chain can have access to real-time data, ensuring that everyone is on the same page and reducing the risk of fraud or error. The immutability of blockchain also ensures that the data cannot be tampered with or manipulated, which is crucial for maintaining the integrity of the supply chain. Another key advantage of blockchain is that it enables the automation of certain processes using smart contracts. These contracts can be programmed to execute automatically when certain conditions are met, reducing the need for intermediaries and speeding up the supply chain process. This can help organizations to streamline their operations and reduce costs, while also improving the overall customer experience.

In addition to these benefits, blockchain technology can also enable organizations to implement more sustainable and ethical practices in their supply chain operations. By providing end-to-end visibility, organizations can ensure that their products are being produced and transported responsibly and sustainably manner, which can be a key differentiator in today's competitive market. The use of blockchain technology in the supply chain industry is still in its early stages, but the potential benefits are significant. As more organizations adopt this technology, we can expect to see greater transparency, efficiency, and trust in the supply chain ecosystem, which will ultimately benefit all participants involved.

According to IBM Blockchain Transparent Supply Solution Brief (2022b), 4 billion units of drugs are supplied through the supply chain every year in the US, which translates into 33–55 million supply chain transactions per day. Using blockchain technology, organizations can use consumer engagement, freshness management, inventory optimization, and product authentication for consumer goods. Elements such as environmental regulations compliance, reusable asset management, counterfeit and fraud detection and reduction prevention, and quality control through the manufacturing process can realize efficiencies.

In the pharmaceutical industry, some industry challenges are damage, contamination, and temperature control which can adversely affect the product. Supply disruptions occur due to theft, fraud, and integrity issues and regulations create an extra cost for compliance. Supply chain adoption in the pharmaceutical industry can fight drug trafficking, fraud detection, cold chain monitoring, transaction settlement, and chargeback reconciliation.

IBM core modules: trace module, documents module, data module, and data subscription module. The Trace module provides tracking of the location and status of material upstream and downstream across the supply chain. The document module helps digitize, manage, and share documentation and certification with authorized users. The data module is the organization's own data storage. The data subscription module is geared towards data analytics to assist in analysis and inventory control. Solutions such as IBM offer organizations the opportunity to customize the blockchain according to their parameters; some extra components that can be helpful are consumer mode, insights mode, customer smart contracts, and custom apps and integration. Consumer mode connects the end user with permissions data regarding the product's origin and quality. The insight module provides access to real-time information about supply chain events and transactions. Custom smart contracts help in establishing off-chain connections by using smart contracts to automate and execute business processes. Custom apps and integration create custom capabilities developed particularly for the organization.

Transparent supply provides a flexible solution to maintain low cost and ensures scalability. Data security is an important facet of an organization. By incorporating hardware-based encryption for data that resides in the cloud, a secure methodology can be employed to access the data reliably and securely. Secured blockchain technology can be designed using data from the cloud that can prevent lock-in and provide additional flexible deployment options. The scalability function of the blockchain provides real-time views for millions of transactions to integrated users thereby achieving transparency. The governance model of a blockchain is industry-specific and provides guidelines and defining policy around data integrity and usage. The blockchain ecosystem evolves as the network grows in membership and participation. Business value is derived by ensuring that every member of the network benefits from participation. According to IBM Food Trust, over 200 companies in the food ecosystem are currently sharing data and tracking food through the blockchain. Atea's Seafood Provenance Network is a blockchain network that provides transparency and sustainability traceability in the seafood industry. PharmaPortal is an industry-wide collaborative blockchain technology to ensure the safety and effectiveness of temperature-controlled drugs and cold chain traceability. Farmer Connect is a blockchain-based technology that is used in the coffee industry to connect farmers to consumers to build shared value.

The foundation of blockchain technology facilitates a decentralized, transparent, immutable, and automated system that creates a multitude of applications in various industries. Digital identity is an all-encompassing data solution for an individual. A blockchain-based digital identity platform enables a unified, interoperable, and tamper-proof framework to benefit users and organizations by providing individuals greater sovereignty over their data. The application of blockchain technology in capital markets provides easier, cheaper, and faster access to capital by reducing barriers and enabling peer-to-peer trading in a faster and more transparent manner as settlement of trades, clearing of financial transactions, reducing transaction costs, and decreasing counterparty risk by streamlining auditing and compliance structure.

Decentralized Finance (DeFi) in Supply Chain

Decentralizing means moving to a system that creates a forum for accessibility. Blockchain technology inherently provides a distributed ledger technology that enables decentralization from traditional centralized financial systems and promotes peer-to-peer sharing of data. Blockchain applications in DeFi support a new economic system to achieve financial access, opportunity, and trust.

Global trade and commerce operate as a consortium where recognizing efficacies in operating global supply chains, managing trade financing, and unlocking new business methodologies to create secure digitization for existing documents is favored. Government organizations desire trust, accountability, and responsiveness. Blockchain technology generates high-performing government functions and makes transactions secure, agile, and cost-effective. There is increased productivity loss in law due to manual operations costs. Blockchain technology provides accessibility, transparency, cost savings, speed, and data integrity. Blockchain healthcare solutions create faster efficient, and more secure medical data management and tracking of medical supplies. Incorporating blockchain-based technology can significantly improve patient care and improve the tracking and circulation of prescription drugs in the global markets. In addition, organizations battle with poor supply chain management that is inefficient and lacks transparency to oftentimes promote exploitation. Blockchain-based supply chain management platforms can provide accurate asset tracking and transparency and facilitate important data collection from sourcing of materials to point of consumption.

Blockchain technology can play a significant role in improving global trade and commerce, government functions, healthcare, and supply chain management. Blockchain technology's features, such as transparency, security, immutability, and efficiency, can bring benefits to these areas. In global trade and commerce, blockchain-based platforms can provide better supply chain visibility, reduce paperwork and transaction costs, improve trade finance, and enhance trust and accountability among trading partners. For example, blockchain-based trade finance platforms can reduce the risk of fraud, streamline payment processing, and facilitate cross-border transactions.

In government functions, blockchain technology can help improve transparency, reduce corruption, and increase efficiency. For instance, blockchain-based voting systems can enhance the integrity and security of the electoral process, while blockchain-based identity management systems can provide secure and tamper-proof identity verification. In healthcare, blockchain-based solutions can improve data interoperability, enhance patient privacy, and facilitate the secure sharing of medical records. For example, blockchain-based medical record systems can enable patients to have more control over their medical data and facilitate the seamless sharing of medical information across healthcare providers.

In supply chain management, blockchain technology can help increase transparency, reduce inefficiencies, and enhance accountability. For instance, blockchain-based supply chain platforms can provide end-to-end visibility into the supply chain, enable real-time tracking of goods, and improve the traceability of products from the source of raw materials to the point of consumption. This can help reduce waste, prevent counterfeiting and fraud, and promote sustainable and ethical practices in the supply chain.

Overall, blockchain technology has the potential to transform various industries and bring benefits such as increased efficiency, transparency, security, and trust. However, there are also challenges and limitations to consider, such as regulatory and legal issues, interoperability, scalability, and adoption barriers.

Blockchain technology underpins macroeconomic and global market growth by providing next-generation solutions that achieve interoperability in exchanging transactional information. The use of blockchain technology in supply chain management creates transparency from the source point to the consumer, accurate tracking of assets, and superior licensing of products, services, and software. Every business generation to this point has created value in business models using technology. Even though the business has advanced and evolved, supply chains remain ineffective in tracking and expediting exploitative behavior due to a lack of transparency. International shipping is manually intensive, and more than half the cost of transport goes to pay for paperwork. The International Ocean Advocacy Organization (OCEANA) conducted a nationwide survey in the U.S. from 2010 to 2012 that revealed up to 87% of seafood is mislabeled. These statistics substantiate the ineffectiveness of current supply chains. Consumption of seafood that is inappropriately procured and handled can cause serious consequences for the consumer (McIlgorm et al., 2022). Food safety in raw foods can create biologically evolved diseases that may not have a cure. Therefore, the food safety of raw foods like seafood needs to urgently evolve to create data transparency and tracking of the supply chain to protect consumers. The implementation of blockchain-based technology will bring traceability, transparency, and accountability to the movement of goods. In addition, blockchain creates efficiency and cost savings for the supply chain infrastructure.

In a typical industry supply chain, there are many complex components such as suppliers, manufacturers, distributors, retailers, auditors, and consumers. Regardless of how large and how many elements are involved in a supply chain, blockchain technology provides an infrastructure that would streamline workflows for participants. The shared distributed ledger creates data security and provides greater visibility to auditors and customers to drive value. Blockchain technology offers improvements by cutting costs from the supply chain infrastructure. Supply chain management can benefit from features such as traceability, transparency, and traceability that are all at the core of blockchain technology.

Blockchain technology has the potential to revolutionize supply chain management by enabling secure, transparent, and efficient tracking of products and transactions across the supply chain. With blockchain, all participants in the supply chain can have access to the same immutable, tamper-proof ledger, reducing the need for intermediaries and improving trust and transparency. One of the key benefits of blockchain technology in supply chain management is the ability to provide end-to-end traceability of products. By recording each step in the supply chain on the blockchain, from raw materials to finished goods, it becomes possible to track the origin of products, verify their authenticity, and ensure compliance with regulations and industry standards. This is particularly important for industries where safety and quality are critical, such as food and pharmaceuticals.

Another benefit of blockchain technology is improved efficiency and cost savings. By eliminating the need for intermediaries and reducing paperwork and manual processes, blockchain can streamline supply chain workflows, reduce transaction costs, and accelerate the movement of goods. This can benefit all participants in the supply chain, from suppliers to consumers, by lowering costs and improving speed and reliability. Blockchain technology has the probability to transform supply chain management by providing greater transparency, traceability, and efficiency. As technology continues to mature and become more widely adopted, we will likely see significant changes in the way that supply chains are managed and operated.

Traceability is directly correlated to operational efficiency. Blockchain maps and visualizes enterprise supply chains by providing consumers who demand sourcing information about the products that they purchase. Blockchain technology creates competitive advantages for organizations that seek to understand and galvanize consumer engagement with real-time, verifiable, and immutable data. A global supply chain ranges from consumer goods to product recalls. In many past instances, consumer products, especially raw ingredients may need to be recalled preventing illness or injury to the consumer. All factors involved in a recall such as lost sales, the replacement cost of the product, and lawsuits harm the organization's reputation, and everyone involved in the production of the product that is being recalled. Blockchain technology provides enhanced product traceability to target the reduction of counterfeiting and streamline the cumbersome traditional product recall system. Counterfeiting costs organizations billions of dollars in losses and reputational losses. From consumer goods to prescription drugs there are billions of dollars at stake. Blockchain provides consumers with real-time, valid, and verified data about how the product was sourced to prevent counterfeiting measures.

Transparency creates trust by recording and sharing key data points such as certification and claims to all users on the blockchain. Blockchain offers real-time validation and verification from third-party attesters to create authenticity. Supply chain traceability is one of the strongest suits of blockchain technology. Refurbishing data from a traditional process and incorporating it in a distributed ledger technology provides blockchain users with the ability to track any digital or physical product throughout its lifecycle. On a global scale, distributed ledger technology can encourage sustainable and ethical production and consumption of products. Products are repurposed and relabeled for use by organizations. Repackaging and relabeling are common practices. Blockchain provides transparency in tracking all source data regarding the product irrespective of repackaging and relabeling and the authenticity of the product remains validated. Smart contracts are used to enforce the asset-tracking process in a blockchain. All users on the blockchain have visibility of the entire journey of the asset in real time.

Tradability is a unique function of blockchain technology to provide a new source to redefine the conventional marketplace. The digital representation of ownership creates value for all shareholders who can transfer ownership without the physical asset changing hands. Verification of past ownership through standardized licensing procedures is essential for many industries. Blockchain facilitates efficient ownership of data and licensing. Using smart contracts, blockchain users can accurately utilize license services, products, and software. The consensus mechanism of blockchain technology creates a no-dispute environment whereby by design transactions are recorded on the distributed ledger, shared with all users, and provided the unique potential to track ownership of records.

Blockchain-based technology enhances regulatory and compliance reporting by reducing friction, reporting costs, and eliminating errors that are directly associated with manual activities. Regulation and compliance present severe concerns for organizations. Organizations earmark billions of dollars to meet regulations and stay compliant. These organizational efforts to maintain safety and regulatory compliance are often manual and cost intensive. The distributed ledger technology at the heart of blockchain enhances governance by providing real-time data to all stakeholders.

Early adopters of blockchain technology believe that blockchain has the potential to unlock productivity, effectiveness, and profitability across businesses that embrace this new wave of technology. While adoption is unwaveringly supported, early adopters forewarn organizations that ignore blockchain technology may be in jeopardy. Since blockchain is a novel technology, executives do not have a clear understanding of what blockchain technology is all about. Therefore, organization leaders often disregard the technology merely due to its complexity and layers that make it rigid to understand.

Application

Blockchain SWOT Analysis: SWOT analysis is a popular strategic planning tool used by businesses, organizations, and individuals to assess their current situation and plan for the future. SWOT stands for Strengths, Weaknesses, Opportunities, and Threats, and the analysis involves identifying and examining each of these four factors (see Figure 12). Here are some reasons why SWOT analysis is useful:

- **It provides a comprehensive overview:** SWOT analysis helps you get a holistic view of your current situation, considering both internal and external factors. By examining your strengths, weaknesses, opportunities, and threats, you can gain a better understanding of where you stand and what you need to do to improve.
- **It helps identify strategic options:** Once you have identified your SWOT factors, you can use them to develop strategies that will help you achieve your goals. For example, you can leverage your strengths to take advantage of opportunities or address your weaknesses to minimize threats.
- **It promotes better decision-making:** SWOT analysis provides a framework for making informed decisions. By considering all four factors, you can weigh the pros and cons of different options and make decisions based on evidence rather than intuition or guesswork.
- **It encourages collaboration:** SWOT analysis is often done as a team exercise, which encourages collaboration and shared ownership of the outcome. By involving multiple stakeholders, you can get a diversity of perspectives and insights that can help you make better decisions.

Therefore, SWOT analysis is a useful tool for gaining a deeper understanding of your situation and developing effective strategies to achieve your goals. Here is a framework that shows the SWOT analysis for blockchain implementation.

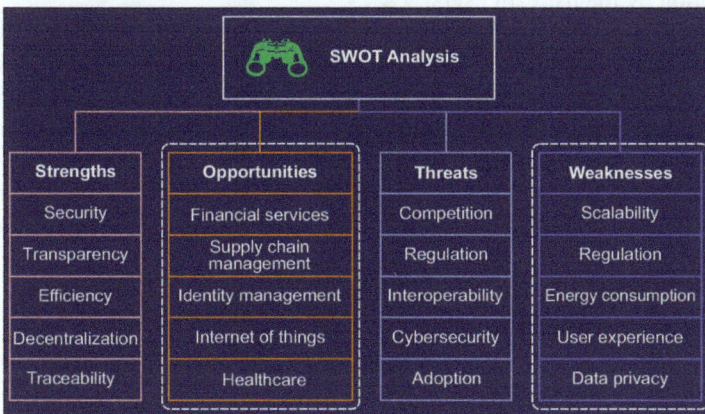

Figure 12: Blockchain SWOT Analysis (Source: Author's own work)

Strengths

1. **Security:** Blockchain is inherently secure due to its decentralized architecture and cryptographic algorithms, making it a reliable platform for sensitive data and financial transactions.
2. **Transparency:** The distributed ledger system allows for transparent and verifiable transactions, increasing trust and reducing the need for intermediaries.
3. **Efficiency:** Blockchain technology can significantly reduce transaction times and costs compared to traditional methods, making it a cost-effective solution for businesses.
4. **Decentralization:** The decentralized nature of blockchain reduces the risk of a single point of failure, making it more resilient to cyberattacks and other disruptions.
5. **Traceability:** The blockchain ledger provides a complete record of all transactions, enabling traceability and accountability in supply chains and other applications.

Weaknesses

1. **Scalability:** The current architecture of blockchain technology limits its scalability, leading to slow transaction speeds and increased processing costs.
2. **Regulation:** The lack of clear regulatory frameworks for blockchain technology creates uncertainty and inhibits adoption by businesses and governments.
3. **Energy Consumption:** The proof-of-work consensus mechanism used in many blockchain networks consumes a significant amount of energy, leading to concerns about environmental sustainability.
4. **User Experience:** The user experience of blockchain technology can be complex and difficult for non-technical users to understand, hindering its adoption by mainstream users.
5. **Data Privacy:** While blockchain is secure, it may not be the best solution for applications requiring strict data privacy and confidentiality, as data on the ledger is visible to all participants. Additionally, the vulnerability of blockchain to hacks from quantum computers is a concern, as this emerging technology has the potential to undermine the cryptographic algorithms that underpin blockchain's security. This could result in the compromise of sensitive data and financial transactions, undermining trust in the technology.

Opportunities

1. **Financial Services:** Blockchain technology can potentially disrupt the financial services industry by reducing costs, improving efficiency, and enabling new business models.
2. **Supply Chain Management:** Blockchain can improve supply chain

transparency, traceability, and accountability, enabling more sustainable and ethical business practices.

3. **Identity Management:** Blockchain can provide a secure and decentralized solution for identity management, improving privacy and reducing the risk of identity theft.

4. **Internet of Things (IoT):** Blockchain can enable secure and efficient communication and transactions between IoT devices, enabling new business models and applications.

5. **Healthcare:** Blockchain technology can revolutionize healthcare by improving patient data management, medical recordkeeping, and clinical trial data sharing. This can lead to better patient outcomes, improved data security, and reduced healthcare costs. For example, blockchain can enable secure and decentralized access to medical records, reducing the risk of data breaches and improving patient privacy. Additionally, blockchain can facilitate the sharing of medical data between different healthcare providers, enabling better collaboration and more informed decision-making.

Threats

1. **Competition:** There are many alternative technologies and solutions that could compete with blockchains, such as traditional databases, cloud-based solutions, and other emerging technologies.

2. Unclear or restrictive regulatory frameworks can inhibit adoption and innovation in the blockchain industry.

3. **Interoperability:** The lack of interoperability between different blockchain networks and protocols can hinder adoption and limit the potential benefits of the technology.

4. **Cybersecurity:** While blockchain is inherently secure, it is not immune to cyberattacks and other security threats, which could undermine trust in the technology.

5. **Adoption:** Despite the potential benefits of blockchain technology, adoption may be slow due to a lack of awareness, understanding, and infrastructure, as well as resistance from established players in certain industries.

Conclusion

IoT, or the Internet of Things, refers to the network of physical devices, vehicles, home appliances, and other items embedded with electronics, software, sensors, and connectivity, allowing them to connect and exchange data with other devices and systems over the internet. The massive amounts of data generated by IoT devices offer valuable insights and opportunities for businesses, governments, and individuals to improve their processes, reduce costs, and enhance their services. For example, in the healthcare industry, IoT-enabled devices can track patient health data and provide real-time monitoring, leading to improved patient outcomes and reduced costs. In the energy industry, IoT sensors can monitor and optimize energy usage in homes and buildings, reducing waste and saving money.

To effectively leverage IoT data, organizations need to have robust data processing and analytics capabilities, including machine learning and artificial intelligence algorithms. This allows them to derive insights from the data, identify patterns and anomalies, and make informed decisions about how to optimize their operations and services. Overall, the IoT offers tremendous potential to improve the quality of life for individuals and society. By harnessing the power of IoT data, we can create more efficient and sustainable systems and improve the way we live, work, and interact with the world around us.

Hyperledger Fabric is a blockchain platform designed to support distributed ledger technology for enterprise applications. One of the key features of Hyperledger Fabric is its modular and configurable architecture, which allows for the integration of multiple modules and components to create a customized blockchain network that meets specific business requirements. This modular architecture also enables Fabric to support multiple consensus algorithms, smart contract languages, and other components that can be tailored to the needs of the network participants. Thus, this makes Hyperledger Fabric a flexible and scalable platform adaptable to a wide range of use cases and business needs.

SWOT analysis is a valuable tool for analyzing the strengths, weaknesses, opportunities, and threats related to a particular situation, such as a business or personal goal. By identifying these factors, individuals or organizations can develop a better understanding of their current situation and use this insight to develop effective strategies for achieving their objectives. The analysis can help individuals or organizations to capitalize on their strengths, address their weaknesses, take advantage of available opportunities, and mitigate potential threats, which ultimately can lead to greater success and achievement of their goals.

Chapter Summary

Privacy concerns have become increasingly important with the rise of data tracking and pattern development. To address these concerns, privacy-enhancing technologies (PETs) have been developed to help minimize privacy risks while still allowing for efficient data management and distribution.

One such technology is differential privacy, which adds noise to data to protect individual privacy while still allowing for accurate statistical analysis. Another is homomorphic encryption, which allows for the computation of encrypted data without the need for decryption, thus minimizing the risk of exposing sensitive information. However, there are limitations to these technologies. For example, differential privacy can decrease the accuracy of analysis, and homomorphic encryption can be computationally expensive. In addition, PETs often require specialized knowledge and resources to implement and maintain, which can be a barrier to adoption for many organizations. Moreover, while PETs can help to address privacy concerns, it's important to recognize their limitations and consider them as part of a broader privacy strategy that also includes policies, procedures, and employee training.

SWOT analysis is a useful tool for assessing the current state of a business or organization and identifying potential areas for improvement or growth. Specifically, it can be valuable in assessing the strengths and weaknesses of the organization, as well as identifying potential opportunities and threats in the external environment. For a blockchain technology company or project, here are some potential areas to consider in a SWOT analysis:

Strengths:

- **Security:** Blockchain technology is often lauded for its security and ability to prevent fraud or tampering.
- **Decentralization:** Blockchain networks are decentralized, meaning that no single entity has control over the data or transactions.
- **Transparency:** Blockchain networks are transparent, meaning that all participants can see the data and transactions on the network.
- **Efficiency:** Blockchain networks can often process transactions more quickly and with lower fees than traditional systems.

Weaknesses:

- **Complexity:** Blockchain technology can be difficult to understand and implement, particularly for non-technical users.
- **Scalability:** Some blockchain networks struggle to handle large volumes of transactions, which can impact their effectiveness.
- **Regulation:** The regulatory environment around blockchain technology is still uncertain in many countries, which can make it difficult for companies to operate or innovate.

Opportunities:

- **New use cases:** Blockchain technology is still a relatively new field, and there may be many untapped use cases for the technology that companies can explore.
- **Partnerships:** Blockchain technology has the potential to facilitate new partnerships between companies or organizations.
- **Efficiency gains:** Blockchain technology has the potential to streamline many business processes, which could lead to significant cost savings.

Threats:

- **Competition:** There are many companies and projects in the blockchain space, which could make it difficult for new entrants to succeed.
- **Security risks:** Despite the inherent security of blockchain technology, there are still risks of hacking or other security breaches.
- **Lack of adoption:** While blockchain technology has received a lot of hype,

it still has relatively low adoption rates in many industries, which could limit its growth potential.

Take Away Lessons from the Chapter

1. Data tracking and pattern development, although not illegal, create privacy concerns for the consumer. To address privacy concerns and to create efficiencies in data management and distribution, a technology that creates specific types of security parameters is required to minimize privacy risks.
2. Blockchain technology platforms address privacy concerns and create utility. In the data sanitizing process, blockchain application facilitates a framework that creates checks and balances while creating visibility of shared data to authorized users.
3. Compared to traditional data storage, blockchain offers an innovative solution to drive the optimization and promotion of transactions in all sectors of industries. The digital era has been gaining momentum from the adoption of massive online transactions by consumers. The convenience of digital technology has catapulted digital services and radically changed economies globally.
4. The rapid evolution of global business architecture creates various opportunities and challenges. As global industries continue to grow, competition and the race for resources continue to demand robust solutions for complex business needs.
5. Since blockchain is a novel technology, executives do not have a clear understanding of what blockchain technology is all about. Therefore, organization leaders often disregard the technology merely due to its complexity and layers that make it rigid to understand.
6. Advanced information technologies such as artificial intelligence (AI) and blockchains have the capacity to ensure efficacy, regulate information asymmetry, and increase trust.
7. SWOT analysis is a popular strategic planning tool used by businesses, organizations, and individuals to assess their current situation and plan. SWOT stands for Strengths, Weaknesses, Opportunities, and Threats, and the analysis involves identifying and examining each of these four factors.

Chapter Exercises

1. What are some of the data privacy concerns that currently exist? How do you think blockchain can help with data privacy issues?
2. Compare traditional databases with blockchain. What are the advantages of blockchain technology and how can blockchain technology evolve old legacy systems?
3. What are the benefits of using AI and blockchain technology?

References

Afzal, M., Huang, Q., Amin,W., Umer, K., Raza, A. and Naeem, M. (2020). Blockchain enabled distributed demand side management in community energy system with smart homes. *IEEE Access*, 8, 37428–37439. https://doi.org/10.1109/ ACCESS.2020.2975233

Agi, M.A.N. and Jha, A.K. (2022). Blockchain technology in the supply chain: An integrated theoretical perspective of organizational adoption. *International Journal of Production Economics*, 247, 108458. Elsevier BV. https://doi.org/10.1016/j. ijpe.2022.108458

Agung, A.A.G. and Handayani, R. (2022). Blockchain for smart grid. *Journal of King Saud University – Computer and Information Sciences*, 34(3), 666–675. https://doi. org/10.1016/j.jksuci.2020.01.002

Akyildirim, E., Corbet, S., Cumming, D., Lucey, B. and Sensoy, A. (2020). Riding the Wave of crypto-exuberance: The potential misusage of corporate blockchain announcements. *Technological Forecasting and Social Change*, 159, 120191. Elsevier BV. https://doi.org/10.1016/j.techfore.2020.120191

Albrecht, C., Duffin, K.M., Hawkins, S. and Morales Rocha, V.M. (2019). The use of cryptocurrencies in the money laundering process. *Journal of Money Laundering Control*, 22(2), 210–216. https://doi.org/10.1108/jmlc-12-2017-0074

Alfandre, D., Stream, S. and Geppert, C. (2020). "Doc, I'm going for a walk": Liberalizing or restricting the movement of hospitalized patients—ethical, legal, and clinical considerations. *HEC Forum*, 32(3), 253–267. https://dx.doi.org/10.1007/s10730-020-09398-5

Alfarsi, F., Lemke, F. and Yang, Y. (2019). The importance of supply chain resilience: An empirical investigation. *Procedia Manufacturing*, 39, 1525–1529. https://doi. org/10.1016/j.promfg.2020.01.295

Alfieri, C. (2022). Cryptocurrency and national security. *International Journal on Criminology*, 9(1). https://doi.org/10.18278/ijc.9.1.3

Algarni, A.D., El Banby, G., Ismail, S., El-Shafai, W., El-Samie, F.E.A. and Soliman, N.F. (2020). Discrete transforms and matrix rotation based cancelable face and fingerprint recognition for biometric security applications. *Entropy*, 22(12), 1361. MDPI AG. https://doi.org/10.3390/e22121361

Ali, O., Ally, M. and Dwivedi, Y. (2020). +e state of play of blockchain technology in the financial services sector: A systematic literature review. *International Journal of Information Management*, 54, Article ID 102199.

Alkhateeb, A., Catal, C., Kar, G. and Mishra, A. (2022). Hybrid blockchain platforms for the Internet of Things (IoT): A systematic literature review. *Sensors*, 22(4), 1304. MDPI AG. https://doi.org/10.3390/s22041304

Allaoui, H., Guo, Y. and Sarkis, J. (2019). Decision support for collaboration planning in sustainable supply chains. *Journal of Cleaner Production*, 229, 761–774. https://doi.org/10.1016/j.jclepro.2019.04.367

Allayannis, G., Fernstrom, A. and Luzgina, A.N. (2018). An introduction to blockchain. *SSRN Electronic Journal*. Elsevier BV. https://doi.org/10.2139/ssrn.3246240

Alletto, S., Cucchiara, R., Del Fiore, G., Mainetti, L., Mighali, V., Patrono, L. and Serra, G. (2016). An indoor location-aware system for an IoT-based intelligent museum. *IEEE Internet Things J*, 3(2), 244–253.

Ambulkar, S., Blackhurst, J.V. and Cantor, D.E. (2016). Supply chain risk mitigation competency: An individual-level knowledge-based perspective. *International Journal of Production Research*, 54(5), 1398–1411. https://doi.org/10.1080/00207543.2015.1070972

Andoni, M., Robu, V., Flynn, D., Abram, S., Geach, D., Jenkins, D., McCallum, P. and Peacock, A. (2019). Blockchain Technology in the Energy Sector: A Systematic Review of Challenges and Opportunities. Elseiver. ScienceDirect. https://doi.org/10.1016/j.rser.2018.10.014.

Angraal, S., Krumholz, H.M. and Schulz, W.L. (2017). Blockchain technology: Applications in health care. *Circulation: Cardiovascular Quality and Outcomes*, 10(9), e003800.

Annual Training Conference (2022). Nhcaa.org. https://www.nhcaa.org/tools-insights/about-health-care-fraud/the-challenge-of-health-care-fraud/

Ansari, A.A., Mishra, B., Gera, P. and Khan, M.K. (2022). Privacy-enabling framework for the cloud-assisted digital healthcare industry. *IEEE Transactions on Industrial Informatics*, 1. https://doi.org/10.1109/TII.2022.3170148

Antes, A.L., Burrous, S., Sisk, B.A., Schuelke, M.J., Keune, J.D. and DuBois, J.M. (2021). Exploring perceptions of healthcare technologies enabled by artificial intelligence: An 102 online, scenario-based survey. *BMC Medical Informatics and Decision Making*, 21, 1–15. https://dx.doi.org/10.1186/s12911-021-01586-8

Asmussen, J.N., Kristensen, J. and Waehrens, B.V. (2018). Cost estimation accuracy in supply chain design: The role of decision-making complexity and management 100 attention. *International Journal of Physical Distribution & Logistics Management*, 48(10), 995–1019. https://doi.org/10.1108/IJPDLM-07-2018-0268

Asset Management and Tracking. https://www.intelligentedgeusa.com/assetmanagement-tracking, Accessed date: 2/25/2023.

Ataseven, C., Nair, A. and Ferguson, M. (2018). An examination of the relationship between intellectual capital and supply chain integration in humanitarian aid organizations: A survey-based investigation of food banks. *Decision Sciences*, 49(5), 827–862. https://doi.org/10.1111/deci.12300

Ateetanan, P. and Shirahada, K. (2018). Factors influencing the adoption of electronic road mapping. *Academy of Strategic Management Journal*, 17(4), 1–13. FINCEN (2023) https://www.fincen.gov/news/news-releases/fincen-identifies-virtual-currency-exchange-bitzlato-primary-money-laundering

Attia, A. and Eldin, I.E. (2018). Organizational learning, knowledge management capability and supply chain management practices in the Saudi food industry. *Journal of Knowledge Management*, 22(6), 1217–1242. https://doi.org/10.1108/JKM-09-2017-0409

Australian Securities and Investments Commission (2022). Crypto-assets. Australian Securities & Investments Commission. https://asic.gov.au/regulatory-resources/digitaltransformation/crypto-assets

Azaria, A., Ekblaw, A., Vieira, T. and Lippman, A. (2016). *MedRec: Using Blockchain for Medical Data Access and Permission Management.* 2016 2nd International Conference on Open and Big Data (OBD), 25–30. https://doi.org/10.1109/obd.2016.11

Back, A., Corallo, M., Dashjr, L., Friedenbach, M., Maxwell, G., Miller, A., Poelstra, A., Timon, J. and Wuille, P. (2014). Enabling blockchain innovations with pegged sidechains. https://blockstream.comtechnology/sidechains

Bak, O., Jordan, C. and Midgley, J. (2019). The adoption of soft skills in supply chain and understanding their current role in supply chain management skills agenda: A UK perspective. *Benchmarking: An International Journal*, 26(3), 1063–1079. https://doi.org/10.1108/BIJ-05-2018-0118

Balon, B., Kalinowski, K. and Paprocka, I. (2022). *Application of Blockchain Technology in Production Scheduling and Management of Human Resources Competencies.* MDPI AG. https://doi.org/10.3390/s22082844

Bals, L., Schulze, H., Kelly, S. and Stek, K. (2019). Purchasing and supply management (PSM) competencies: Current and future requirements. *Journal of Purchasing and Supply Management*, 25(5), 100572. https://doi.org/10.1016/j.pursup.2019.100572

Bamakan, S.M.H., Malekinejad, P. and Ziaeian, M. (2022). Towards blockchain-based hospital waste management systems; applications and future trends. *Journal of Cleaner Production*, 349, 131440. Elsevier BV. https://doi.org/10.1016/j.jclepro.2022.131440

Banach, R. (2020). *Blockchain Applications Beyond the Cryptocurrency Casino: The Punishment not Reward Blockchain Architecture.* Wiley. https://doi.org/10.1002/cpe.5749

Bani, I.W., Akour, A., Ibrahim, A., Almarzouqi, A., Abbas, S., Hisham, F. and Griffiths, J. (2020). Privacy, confidentiality, security, and patient safety concerns about electronic health records. *International Nursing Review*, 67(2), 218–230. https://doi.org/10.1111/inr.12585

Barakat, M.T., Mithal, A., Huang, R.J., Mithal, A., Sehgal, A., Banerjee, S. and Singh, G. (2017). Affordable care act and healthcare delivery: A comparison of California and Florida hospitals and emergency departments [Report]. *PLoS One*, 12, e0182346.

Barmes, I., Kohn, I. and Soutar, C. (2022, April 29). *What does the Dawn of Quantum Computing mean for Blockchain?* World Economic Forum. https://www.weforum.org/agenda/2022/04/could-quantum-computers-steal-the-bitcoins-straight-out-of-your-wallet/

Barone, R. and Masciandaro, D. (2019). Cryptocurrency or usury? Crime and alternative money laundering techniques. *European Journal of Law and Economics*, 47(2), 233–254. https://doi.org/10.1007/s10657-019-09609-6

Bashir, M.R. and Gill, A.Q. (2016). *Towards an IoT Big Data Analytics Framework: Intelligent Buildings Systems, High Performance Computing and Communications.* IEEE 14th International Conference on Intelligent City; IEEE 2nd International Conference on Data Science and Systems, 2016 IEEE 18th International Conference on, 2016, December, pp. 1325–1332.

Berthold, K.R. (2019). Supply chain management: A descriptive conception. *International Journal for Empirical Education and Research*, 3(19), 43–65.

Bessant, J., Kaplinsky, R. and Lamming, R. (2003). Putting supply chain learning into practice. *International Journal of Operations & Production Management*, 23(2), 167–184. https://doi.org/10.1108/01443570310458438

Bevc, C.A., Retrum, J.H. and Varda, D.M. (2015). New perspectives on the "Silo Effect": Initial comparisons of network structures across health collaboratives. *American Journal of Public Health*, 105(Suppl 2), S230–S235. https://doi.org/10.2105/AJPH.2014.302256

Bigini, G., Freschi, V. and Lattanzi, E. (2020). A review on Blockchain for the Internet of Medical Things: Definitions, challenges, applications, and vision. *Future Internet*, 12(12), 208. MDPI AG. https://doi.org/10.3390/fi12120208

Biotto, M., De Toni, A.F. and Nonino, F. (2012). Knowledge and cultural diffusion along the supply chain as drivers of product quality improvement: The Illycaffè case study. *The International Journal of Logistics Management*, 23(2), 212–237. https://doi.org/10.1108/09574091211265369

Bravo, M.I.R., Moreno, A.R. and Llorens-Montes, F.J. (2016). Supply network-enabled innovations. An analysis based on dependence and complementarity of capabilities. *Supply Chain Management: An International Journal*, 21(5), 642–660. https://doi.org/10.1108/SCM-02-2016-0062

Brauner, P., Philipsen, R., Calero Valdez, A. and Ziefle, M. (2019). What happens when decision support systems fail? The importance of usability on performance in erroneous systems. *Behaviour & Information Technology*, 38(12), 1225–1242. https://doi.org/10.1080/0144929X.2019.1581258

Breu, S. and Seitz, T.G. (2018). Legislative regulations to prevent terrorism and organized crime from using cryptocurrencies and its effect on the economy and society. *Forthcoming, Legal Impact on the Economy: Methods, Results, Perspectives*. https://ssrn.com/abstract=3081911

Breite, R. and Koskinen, K.U. (2014). Supply chain as an autopoietic learning system. *Supply Chain Management: An International Journal*, 19(1), 10–16. https://doi.org/10.1108/SCM-06-2013-0176

Buchholz, K. (2022). *Which Countries Spend the most Time on Social Media?* https://www.weforum.org/agenda/2022/04/social-media-internet-connectivity/

Butler, M. (2018). Ensuring data integrity during health information exchange. *Journal of AHIMA*, 89(10), 14–17.

Butt, A.S. (2021). Determinants of top-down knowledge hiding in firms: An individual-level perspective. *Asian Business & Management*, 20(2), 259–279. https://doi.org/10.1057/s41291-019-00091-1

Cainelli, G., Montresor, S. and Vittucci Marzetti, G. (2012). Production and financial linkages in inter-firm networks: Structural variety, risk-sharing and resilience. *Journal of Evolutionary Economics*, 22(4) 711–734.

Casey, M.J. and Wong, P. (2017). Global supply chains are about to get better, thanks to blockchain. *Harv. Bus. Rev.*, 13, 1–6.

Casino, F., Dasaklis, T.K. and Patsakis, C. (2019). A systematic literature review of blockchain-based applications: Current status, classification and open issues. *Telemat. Inform.*, 36, 55–81. https://doi.org/10.1016/j.tele.2018.11.006

Chang, V., Baudier, P., Zhang, H., Xu, Q., Zhang, J. and Arami, M. (2020). How Blockchain can impact financial services – The overview, challenges and recommendations from expert interviewees. *Technological Forecasting and Social Change*, 158, Article ID 120166.

Chen, Y., Ding, S., Xu, Z., Zheng, H. and Yang, S. (2019). Blockchain-based medical records secure storage and medical service framework. *Journal of Medical Systems*, 43(1), 1. https://dx.doi.org/10.1007/s10916-018-1121-4

Cherrez-Ojeda, I., Vanegas, E., Felix, M., Mata, V.L., Jiménez, F.M., Sanchez, M.,

Simancas-Racines, D., Cherrez, S., Gavilanes, A.W.D., Eschrich, J. and Chedraui, P. (2020). Frequency of use, perceptions and barriers of information and communication technologies among Latin American physicians: An Ecuadorian cross-sectional study. *Journal of Multidisciplinary Healthcare*, 13, 259–269.

Chowdhury, S. (2002). *The Power of Design for Six Sigma*. Dearborn Trade Publishing.

Christidis, K. and Devetsikiotis, M. (2016). *Blockchains and Smart Contracts for the Internet of Things*. Institute of Electrical and Electronics Engineers (IEEE). https://doi.org/10.1109/access.2016.2566339

Christie's Annual Report (2019). *Annual Report*. https://www.christies.com/about-us/press-archive/details?PressReleaseID=9583&lid=1

Christie's Annual Report (2021). *Annual Report*. https://www.christies.com/about-us/welcome-to-christies/annual-report/

Clayton, J. (2017, December 11). *Statement on Cryptocurrencies and Initial Coin Offerings*. U.S. Securities and Exchange Commission. https://www.sec.gov/news/publicstatement/ statement-clayton-2017-12-11#

Clohessy, T. and Acton, T. (2019). Investigating the influence of organizational factors on blockchain adoption: An innovation theory perspective. *Ind. Manag. Data Syst.*, 119(7), 1457–1491.

CMS (2022, December 15). *Historical*. CMS.gov. Retrieved March 5, 2023, from https://www.cms.gov/research-statistics-data-and-systems/statistics-trends-and-reports/nationalhealthexpenddata/nationalhealthaccountshistorical

Community Certified Policing Services (2015). *U.S. Department of Justice: The President's Task Force on 21st Century Policing*. https://cops.usdoj.gov/pdf/taskforce/taskforce_finalreport.pdf

Corbet, S., Meegan, A., Larkin, C., Lucey, B. and Yarovaya, L. (2018). Exploring the dynamic relationships between cryptocurrencies and other financial assets. *Economics Letters*, 165, 28–34. Elsevier BV. https://doi.org/10.1016/j.econlet.2018.01.004

Cormack, A., Thomé, A.M.T. and Silvestre, B. (2021). An integrative conceptual framework for supply chain sustainability learning: A process-based approach. *Journal of Cleaner Production*, 320, 128675. https://doi.org/10.1016/j.jclepro.2021.128675

Crosby, M., Pattanayak, P., Verma, S. and Kalyanaraman, V. (2016). Blockchain technology: Beyond bitcoin. *Applied Innovation*, 2(6–10), 71.

Dahan, M. and Gelb, A. (2015). The role of identification in the post-2015 development agenda. World Bank. http://hdl.handle.net/10986/22513

Da Lomba, T.G. (2022). Artist's statement: The doctor will see you now. *Academic Medicine*, 97(6), 847. https://doi.org/10.1097/acm.0000000000004655

Dagher, G.G., Mohler, J., Milojkovic, M. and Marella, P.B. (2018). Ancile: Privacy-preserving framework for access control and interoperability of electronic health records using blockchain technology. *Sustainable Cities and Society*, 39, 283–297.

Daly, L. (2021). *What is Byzantine Fault Tolerance? | The Motley Fool*. The Motley Fool. https://www.fool.com/investing/stock-market/market-sectors/financials/cryptocurrency-stocks/byzantine-fault-tolerance/

Department of Justice (2022). *Two Arrested for Alleged Conspiracy to Launder $4.5 billion in Stolen Cryptocurrency*. https://www.justice.gov/opa/pr/two-arrested-alleged-conspiracy-launder-45-billion-stolen-cryptocurrency

de Vocht, F. and Röösli, M. (2021). Comment on Choi, Y.-J., et al. Cellular phone use and risk of tumors: Systematic review and meta-analysis. *Int. J. Environ. Res. Public Health* 2020, 17, 8079. *International Journal of Environmental Research and Public Health*, 18(6). https://doi.org/10.3390/ijerph18063125

De, N. (2022, August 3). *CFTC would Become Primary Crypto Regulator Under New Senate Committee Plan*. CoinDesk: Bitcoin, Ethereum, Crypto News and Price Data. https://www.coindesk.com/policy/2022/08/03/cftc-would-become-primary-cryptoregulator-under-new-senate-committee-plan/

del Rosario Pérez-Salazar, M., Aguilar Lasserre, A.A., Cedillo-Campos, M.G. and Hernández González, J.C. (2017). The role of knowledge management in supply chain management: A literature review. *Journal of industrial Engineering and Management*, 10(4), 711–788. http://dx.doi.org/10.3926/jiem.2144

Derwik, P. and Hellstrom, D. (2017). Competence in supply chain management: A systematic review. *Supply Chain Management*, 22(2), 200–218. http://dx.doi.org/10.1108/SCM-09-2016-0324

Derwik, P. and Hellstrom, D. (2021). How supply chain professionals learn at work: An investigation of learning mechanisms. *International Journal of Physical Distribution & Logistics Management*, 51(7), 738–863. https://doi.org/10.1108/IJPDLM-11-2019-0335

Di Silvestre, M.L., Gallo, P., Guerrero, J.M., Musca, R., Sanseverino, E.R., Sciumè, G., Vasquez, J.C. and Zizzo, G. (2020). Blockchain for power systems: Current trends and future applications. *Renew. Sustainable Energy Reviews*, 119. https://doi.org/10.1016/j.rser.2019.109585

Dierksmeier, C. and Seele, P. (2016). Cryptocurrencies and business ethics. *Journal of Business Ethics*, 152(1), 1–14. https://doi.org/10.1007/s10551-016-3298-0

Du, J., Jiang, C., Han, Z., Zhang, H., Mumtaz, S. and Ren, Y. (2019). Contract mechanism and performance analysis for data transaction in mobile social networks. *IEEE Transactions on Network Science and Engineering*, 6(2), 103–115.

Dubovitskaya, A., Xu, Z., Ryu, S., Schumacher, M. and Wang, F. (2017). Secure and trustable electronic medical records sharing using blockchain. *AMIA Annual Symposium Proceedings*, Vol. 2017, p. 650. American Medical Informatics Association.

Duggan, W. (2022). *The History of Bitcoin, the First Cryptocurrency*. U.S. News. https://money.usnews.com/investing/articles/the-history-of-Bitcoin#

Durneva, P., Cousins, K. and Chen, M. (2020). The current state of research, challenges, and future research directions of blockchain technology in patient care: A systematic review. *Journal of Medical Internet Research*, 22(7), e18619.

Dwivedi, A.D., Srivastava, G., Dhar, S. and Singh, R. (2019). A decentralized privacy-preserving healthcare blockchain for IoT [Article]. *Sensors* (14248220), 19(2), 326. https://doi.org/10.3390/s19020326

Dyntu, V. and Dykyi, O. (2019). Cryptocurrency in the system of money laundering. *Baltic Journal of Economic Studies*, 4(5), 75. https://doi.org/10.30525/2256-0742/2018-4-5-75-81

Edmonds (2023). The new world of meta finance and its yet to be tested efficiencies. *Journal of Securities Operations & Custody*, 15(1), 68–81.

Elmountasser, M. (2019). A behavioural risk perspective of supply chain management: A general overview. Revue des Etudes et Recherche en Logistique et Développement, 4, 1–12. https://revues.imist.ma/index.php/RERLED/article/view/15401

Erturk, E., Lopez, D. and Yu, W.Y. (2020). Benefits and risks of using blockchain in smart energy: A literature review. *Contemporary Management Research*, 15(3), 205–225. https://doi.org/10.7903/cmr.19650

Eyal, I., Gencer., A.E., Sirer, E.G. and van Renesse, R. (2016). *Bitcoin-NG: A Scalable Blockchain Protocol*. Proceedings of the 13th USENIX Symposium on Networked Systems Design and Implementation. USENIX Association.

Fahimnia, B., Pournader, M., Siemsen, E., Bendoly, E. and Wang, C. (2019). Behavioral operations and supply chain management – A review and literature mapping. *Decision Sciences*, 50(6), 1127–1183. https://doi.org/10.1111/deci.12369

FINCEN (2023). FinCEN Identifies Virtual Currency Exchange Bitzlato as a "Primary Money Laundering Concern" in Connection with Russian Illicit Finance. https://www.fincen.gov/news/news-releases/fincen-identifies-virtual-currency-exchange-bitzlato-primary-money-laundering

Farouk, A., Alahmadi, A., Ghose, S. and Mashatan, A. (2020). Blockchain platform for industrial health care: Vision and future opportunities. *Computer Communications*, 154, 223–235.

Flint, D.J., Larsson, E. and Gammelgaard, B. (2008). Exploring processes for customer value insights, supply chain learning and innovation: An international study. *Journal of Business Logistics*, 29(1), 257–281. https://doi.org/10.1002/j.2158-1592.2008.tb00078.x

Flitter, E. and Yaffe-Bellany, D. (2022). Russia could use cryptocurrency to blunt the force of U.S. sanctions. *The New York Times*. https://www.nytimes.com/2022/02/23/business/russia-sanctions-cryptocurrency.html

Flothmann, C., Hoberg, K. and Gammelgaard, B. (2018). Disentangling supply chain management competencies and their impact on performance: A knowledge-based view. *International Journal of Physical Distribution & Logistics Management*, 48(6), 630–655. https://doi.org/10.1108/IJPDLM-02-2017-0120

Foti, M. and Vavalis, M. (2019). Blockchain based uniform price double auctions for energy markets. *Applied Energy*, https://doi.org/10.1016/j.apenergy.2019.113604.

Francisco, M.G., Junior, O.C. and Sant'Anna, Â.M.O. (2020). Design for six sigma integrated product development reference model through systematic review. *International Journal of Lean Six Sigma*, 11, 767–795.

Friedlob, G.T. and Plewa, F.J. (1996). *Understanding Balance Sheets*. Wiley.

Garg, N., Inman, J.J. and Mittal, V. (2017). Emotion effects on choice deferral: The moderating role of outcome and process accountability. *European Journal of Marketing*, 51(9/10), 1631–1649. https://doi.org/10.1108/EJM-12-2015-0861

Garvin, E. (2019). Hackers in healthcare: What damage could they do with your medical data? https://hitconsultant.net/2019/03/05/hackers-in-healthcare-medical-data/

George, G., McGahan, A.M. and Prabhu, J. (2012). Innovation for inclusive growth: Towards a theoretical framework and research agenda. *Journal of Management Studies*, 49, 661–683. https://doi.org/10.1111/j.1467-6486.2012.01048.x

Gloet, M. and Samson, D. (2018). *The Role of Knowledge Management In Innovative Supply Chain Design*. Proceedings of the 51st Hawaii International Conference on System Sciences, 4141–4180. http://hdl.handle.net/10125/50413

Gong, Y., Jia, F., Brown, S. and Koh, L. (2018). Supply chain learning of sustainability in multi-tier supply chains: A resource orchestration perspective. *International Journal of Operations & Production Management*, 38(4), 1061–1090. https://doi.org/10.1108/IJOPM-05-2017- 0306

González-Zarzosa, S. and Diaz, R. (2021). Security information and event management (SIEM): Analysis, trends, and usage in critical infrastructures. *Sensors*, 21(14), 4759. https://dx.doi.org/10.3390/s21144759

Gopalakrishnan, P.K., Hall, J. and Behdad, S. (2021). Cost analysis and optimization of blockchain-based solid waste management traceability system. *Waste Management*, 120, 594–607. Elsevier BV. https://doi.org/10.1016/j.wasman.2020.10.027

Goranovic, A., Meisel, M., Wilker, S. and Sauter, T. (2019). *Hyperledger Fabric Smart Grid Communication Testbed on Raspberry PI ARM Architecture*. 15th IEEE International Workshop on Factory Communication Systems (WFCS). https://doi.org/10.1109/WFCS.2019.8758000

Goudarzi, F.S., Bergey, P. and Olaru, D. (2021). Behavioral operations management and supply chain coordination mechanisms: A systematic review and classification of the literature. *Supply Chain Management: An International Journal*. https://doi.org/10.1108/SCM-03-2021-0111

Green, J. and Boyle, M. (2022). What is the law for Cryptocurrency? Regulations are on the Horizon. Trading, Cryptocurrency & Bitcoin. Guide to Cryptocurrency. The Balance. https://www.thebalancemoney.com/what-are-the-laws-for-cryptocurrency-5121102

Gunduz, M.Z. and Das, R. (2020). Cyber-security on smart grid: Threats and potential solutions. *Computer Networks*, 169, 107094.

Guo, Y., Wan, Z. and Cheng, X. (2022). When blockchain meets smart grids: A comprehensive survey. *High-Confidence Computing*, 2(2), Article ID 100059.

Guo, Y., Wang, L., Ma, R., Mu, Q., Yu, N., Zhang, Y., Tang, Y., Li, Y., Jiang, G., Zhao, D., Mo, F., Gao, S., Yang, M., Kan, F., Ma, Q., Fu, M. and Zhang, D. (2016). JiangTang XiaoKe granule attenuates cathepsin K expression and improves IGF-1 expression in the bone of high fat diet induced KK-Ay diabetic mice. *Life Sciences*, 148, 2016. https://doi.org/10.1016/j.lfs.2016.02.056

Gupta, V. and Knight, R. (2017). Economic development: How blockchain could help emerging markets leap ahead. *Harvard Business Review*. https://www.thebalance.com/what-are-the-laws-for-cryptocurrency-5121102

Haines, R., Hough, J. and Haines, D. (2017). A metacognitive perspective on decision making in supply chains: Revisiting the behavioral causes of the bullwhip effect. *International Journal of Production Economics*, 184, 7–20. https://doi.org/10.1016/j.ijpe.2016.11.006

Hale, C., Uglow, S. and Heaton, R. (2005). Uniform styles II: Police families and policing styles. *Policing and Society*, 15(1).

Haleem, A., Javaid, M., Singh, R.P., Suman, R. and Rab, S. (2021). Blockchain technology applications in healthcare: An overview. *International Journal of Intelligent Networks*, 2, 130–139.

Halik, A., Verweij, M. and Schlüter, A. (2018). How marine protected areas are governed: A cultural theory perspective. *Sustainability*, 10(1). https://doi.org/10.3390/su10010252

Haq, M.Z.U. (2020). Supply chain learning and organizational performance: Evidence from Chinese manufacturing firms. *Journal of Knowledge Management*, 25(4), 943–972. https://doi.org/10.1108/JKM-05-2020-0335

Harris, L., Liboiron, M., Charron, L. and Mather, C. (2021). Using citizen science to evaluate extended producer responsibility policy to reduce marine plastic debris shows no reduction in pollution levels. Mar. Policy 123. https://doi.org/10.1016/j.marpol.2020.104319

Harrison, K. and Leopold, A. (2021). *How Blockchain can Help Combat Disinformation*. Harvard Business Review.

Hassan, M.H., Sharifah, M.Y., Nur, I.U. and Ninggal, M.I.H. (2021). Blockchain-based access control scheme for secure shared personal health records over decentralised storage. *Sensors*, 21(7), 2462. https://dx.doi.org/10.3390/s21072462

Herazo-Padilla, N., Augusto, V., Dalmas, B., Xie, X. and Bongue, B. (2021). A decision-tree based bayesian approach for chance-constrained health prevention budget rationing. *IEEE Transactions on Automation Science and Engineering*, 1–17. https://doi.org/10.1109/tase.2021.3069800

Heyler, S.G. and Martin, J.A. (2018). Servant leadership theory: Opportunities for additional theoretical integration. *Journal of Managerial Issues*, 30(2), 230–243. http://www.jstor.org/stable/45176580

Hoffman, C. (2018). *What's the Difference Between a Bitcoin Wallet and an Exchange? How to Geek*. https://www.howtogeek.com/347464/whats-the-difference-between-a-bitcoinwallet- and-an-exchange/

Hofstra, N., Dullaert, W., De Leeuw, S. and Spiliotopoulou, E. (2019). Individual goals and social preferences in operational decisions. *International Journal of Operations & Production Management*, 39(1), 116–137. https://doi.org/10.1108/IJOPM-11-2016-0690

Hoque, S. and Hoque, S. (2018). *Full-Stack React Projects: Modern Web Development Using React 16, Node, Express, and MongoDB*. Packt Publishing Limited.

Howson, P. (2020). *Building Trust and Equity in Marine Conservation and Fisheries Supply Chain Management with Blockchain*. Elsevier BV. https://doi.org/10.1016/j.marpol.2020.103873

Hu, C., Li, C., Zhang, G., Lei, Z., Shah, M., Zhang, Y., Xing, C., Jiang, J. and Bao, R. (2022). CrowdMed-II: A blockchain-based framework for efficient consent management in health data sharing. *World Wide Web*, 25(3), 1489–1515. Springer Science and Business Media LLC. https://doi.org/10.1007/s11280-021-00923-1

Hughes, C. and Brown, R. (2022). Financial investigation for routine policing in Australia. *Trends and Issues in Crime and Criminal Justice*, 649, 1–13. https://doi.org/10.52922/ti78603

Hult, G.T.M., Nichols, E.L., Giunipero, L.C. and Hurley, R.F. (2000). Global organizational learning in the supply chain: A low versus high learning study. *Journal of International Marketing*, 8(3), 61–83. https://doi.org/10.1509/jimk.8.3.61.19628

Hussein, F. and Sweet, K. (2022). *New Crypt Oversight Legislation Arrives as Industry Shakes ALM Law*. https://www.law.co/dailybuinessreview/2022/08/04/new-crpto-oversightlegistlation-arrives-ad-industry-shakes

Iansiti, M. and Lakhani, K.R. (2017). The truth about blockchain. *Harv. Bus. Rev.*, 95.

IBM (2022a). *IBM Blockchain Services for Supply Chain Solution Brief*. https://www.ibm.com/blockchain-supply-chain

IBM. (2022b) *IBM Blockchain Transparent Supply Solution Brief*. https://www.ibm.com/products/supply-chain-intelligence-suite/blockchain-transparent-supply

Innes, M., Fielding, N. and Cope, N. (2005). The appliance of science? The theory and practice of crime intelligence analysis. *British Journal of Criminology*, 45(1), 39–57.

Ivanovski, K. and Hailemariam, A. (2023). Forecasting the stock-cryptocurrency relationship: Evidence from a dynamic GAS model. *International Review of Economics & Finance*, 86, 97–111. Elsevier BV. https://doi.org/10.1016/j.iref.2023.03.008

i-SCOOP (2017). *Innovation and IoT Case: Intelligent Facility Management Services at Heijmans*. https://www.i-scoop.eu/internet-of-things-iot/innovation-iot-case-smart-facility-management/

Janssen, M., Weerakkody, V., Ismagilova, E., Sivarajah, U. and Irani, Z. (2020). A framework for analysing blockchain technology adoption: Integrating institutional, market and technical factors. *International Journal of Information Management*, 50, 302–309. Elsevier BV. https://doi.org/10.1016/j.ijinfomgt.2019.08.012

Jenab, K., Wu, C. and Moslehpour, S. (2018). Design for six sigma: A review. *Management Science Letters*, 8(1), 1–18.

Jia, F., Gong, Y. and Brown, S. (2019). Multi-tier sustainable supply chain management: The role of supply chain leadership. *International Journal of Production Economics*, 217, 44–63. https://doi.org/10.1016/j.ijpe.2018.07.022

Johnson, T. (2021, August 23). Blockchain and law enforcement: A solution for evidence mismanagement. *Harvard Technology Review*. Retrieved February 25, 2023, from https://harvardtechnologyreview.com/2021/08/22/blockchain-and-law-enforcement-a-solution-for-evidence-mismanagement/

Jordan, C. and Bak, O. (2016). The growing scale and scope of the supply chain: A reflection on supply chain graduate skills. *Supply Chain Management: An International Journal*, 21(5), 610–626. http://dx.doi.org/10.1108/SCM-02-2016-0059

Jung, B.R., Choi, K.S. and Lee, C.S. (2022). Dynamics of dark web financial marketplaces: An exploratory study of underground fraud and scam business. *CrimRxiv*. https://doi.org/10.21428/cb6ab371.dbbe560f

Kamau, C.G. (2022, August 26). The cryptocurrency market in Kenya: A review of awareness and participation by the youth. *Journal of Asian Business Strategy*, 12(1), 49–56. https://doi.org/10.55493/5006.v12i1.4599

Kamble, S.S., Gunasekaran, A. and Sharma, R. (2020). Modeling the blockchain enabled traceability in agriculture supply chain. *International Journal of Information Management*, 52, 101967. Elsevier BV. https://doi.org/10.1016/j.ijinfomgt.2019.05.023

Kapoor, K.K., Dwivedi, Y.K. and Williams, M.D. (2014). Rogers' innovation adoption attributes: A systematic review and synthesis of existing research. *Inf. Syst. Manag.*, 31(1), 74–91.

Kassou, M., Bourekkadi, S., Khoulji, S., Slimani, K., Chikri, H. and Kerkeb, M.L. (2021). Blockchain-based medical and water waste management conception. *In:* S. Bourekkadi, J. Abouchabaka, O. Omari and K. Slimani (Eds.), *E3S Web of Conferences*, 234, 00070. EDP Sciences. https://doi.org/10.1051/e3sconf/202123400070

Katarzyna, C. (2019, March). Cryptocurrencies: Opportunities, risks, and challenges for anticorruption compliance systems. OECD Conference on *Global Anti-corruption and Integrity Forum*. Paris (pp. 20–21). https://coincapp.com/corruption/integrityforum/academic-papers/Ciupa-Katarzyna-cryptocurrencies.pdf

Kaur, M., Gupta, S., Kumar, D., Verma, C., Neagu, B.C. and Raboaca, M.S. (2022). Delegated Proof of Accessibility (DPoAC): A novel consensus protocol for blockchain systems. *Mathematics*, 10(13), 2336. MDPI AG. https://doi.org/10.3390/math10132336

Khajeh, H., Laaksonen, H., Gazafroudi, A.S. and Shafie-Khah, M. (2020). Towards flexibility trading at TSO-DSO-customer levels: A review. *Energies*, 13(1), 165. https://doi.org/10.3390/en13010165

Khan, H. and Masood, T. (2022). Impact of blockchain technology on smart grids. *Energies*, 15, 7189. https://doi.org/10.3390/en15197189

Khan, R. and Hakami, T.A. (2021, November 29). Cryptocurrency: Usability perspective versus volatility threat. *Journal of Money and Business*, 2(1), 16–28. https://doi.org/10.1108/jmb-11-2021-0051

Khan, U., An, Z.Y. and Imran, A. (2020). A blockchain Ethereum technology-enabled digital content: Development of trading and sharing economy data. *IEEE Access*, 8, 217045–217056.

Kim, C. and Hyun Jung, K. (2019). A study on healthcare supply chain management efficiency: Using bootstrap data envelopment analysis. *Health Care Management Science*, 22(3), 534–548. https://dx.doi.org/10.1007/s10729-019-09471-7

Kirpes, B., Mengelkamp, E., Schaal, G. and Weinhardt, C. (2019). Design of a microgrid local energy market on a blockchain-based information system. *IT – Information Technology*, 61(2–3), 87–99. https://doi.org/10.1515/itit-2019-0012

Kosajan, V., Chang, M., Xiong, X., Feng, Y. and Wang, S. (2018). *The Design and Application of a Government Environmental Information Disclosure Index in China*. Elsevier BV. https://doi.org/10.1016/j.jclepro.2018.08.056

Kouhizadeh, M., Saberi, S. and Sarkis, J. (2021). Blockchain technology and the sustainable supply chain: Theoretically exploring adoption barriers. *Int. J. Prod. Econ.*, 231, 107831.

Kovvali, A. (2022). Stakeholderism Silo Busting. *SSRN Electronic Journal*. Elsevier BV. https://doi.org/10.2139/ssrn.4240979

Kramer, M. (2019). An overview of Blockchain technology based on a study of public awareness. *Global Journal of Business Research*, 13(1), 83–91. https://ssrn.com/abstract=3381119

Kshetri, N. and Loukoianova, E. (2019). Blockchain adoption in supply chain networks in Asia. *IT Professional*, 21(1), 11–15, Jan.-Feb. 2019, doi: 10.1109/MITP.2018.2881307

Kull, T.J., Oke, A. and Dooley, K.J. (2014). Supplier selection behavior under uncertainty: Contextual and cognitive effects on risk perception and choice. *Decision Sciences*, 45(3), 467–505. https://doi.org/10.1111/deci.12078

Kumar, A.G. and Shantala, C.P. (2020). An extensive research survey on data integrity and deduplication towards privacy in cloud storage. *International Journal of Electrical and Computer Engineering*, 10(2), 2011–2022. https://dx.doi.org/10.11591/ijece.v10i2.pp2011-2022

Kumar, S., Kumar, B. and Nagesh, Y. (2022). Application of blockchain technology as a support tool in economic & financial development. *Manager-3e British Journal of Administrative Management*, 58, 1–14.

Lacity, M.C. (2021). *Fake News, Technology and Ethics: Can AI and Blockchains Restore Integrity?* SAGE Publications. https://doi.org/10.1177/2043886921999065

Lambrechts, F., Taillieu, T., Grieten, S. and Poisquet, J. (2012). In-depth joint supply chain learning: Towards a framework. *Supply Chain Management: An International Journal*, 17(6), 624–637. http://dx.doi.org/10.1108/13598541211269238

Landi, H. (2020). IBM roll outs blockchain network to address supply chain issues caused by COVID-19. https://www.fiercehealthcare.com/tech/ibm-rolls-out-blockchain-network-to-match-healthcare-organizations-non-traditional-suppliers

Larious-Hernandez, G.J. (2017). *Blockchain Entrepreneurship Opportunity in the Practices of the Unbanked*. https://doi.org/10.1016/j.bushor.2017.07.012

Lawal, A.K., Rotter, T., Kinsman, L., Sari, N., Harrison, L., Jeffery, C., Kutz, M., Khan, M.F. and Flynn, R. (2014). Lean management in health care: Definition, concepts, methodology and effects reported (systematic review protocol). *Systematic Reviews*, 3(103), https://doi.org/10.1186/2046-4053-3-103

Lazer, D.M.J., Baum, M.A., Benkler, Y., Berinsky, A.J., Greenhill, K.M., Menczer, F. and Zittrain, J.L. (2018). *The Science of Fake News*. American Association for the Advancement of Science (AAAS). https://doi.org/10.1126/science.aao2998

Lee, J., Moon, S. and Park, J. (2017a). CloudRPS: A cloud analysis based enhanced ransomware prevention system. *Journal of Supercomputing*, 73(7), 3065–3084. https://doi.org/10.1007/s11227-016-1825-5

Lee, S., Lee, N., Ahn, J., Kim, J., Moon, B., Jung, S.H. and Han, D. (2017b). *Construction of an Indoor Positioning System for Home Iot Applications*. 2017 IEEE International Conference on Communications (ICC), May, pp. 1–7 (IEEE).

Lestari, P. and Trihadiningrum, Y. (2019). The impact of improper solid waste management to plastic pollution in Indonesian coast and marine environment. *Mar. Pollut. Bull.*, 149. https://doi.org/10.1016/j.marpolbul.2019.110505

Li, J.O., Liu, H., Ting, D.S.J., Jeon, S., Chan, R.V.P., Kim, J.E., Sim, D.A., Thomas, P.B.M., Lin, H., Chen, Y., Sakomoto, T., Loewenstein, A., Lam, D.S.C., Pasquale, L.R., Wong, T.Y., Lam, L.A. and Ting, D.S.W. (2021). Digital technology, tele-medicine and artificial intelligence in ophthalmology: A global perspective. *Prog. Retin. Eye Res.*, 82, 100900. https://doi.org/10.1016/j.preteyeres.2020.100900

Li, X., Lu, R., Liang, X., Shen, X., Chen, J. and Lin, X. (2011). Intelligent community: An IoT application. *IEEE Commun. Mag*, 49(11).

Li, Z. and Yan, G. (2015). Exploring different order decision behaviors with bullwhip effect and service level measures in supply chain system. *Discrete Dynamics in Nature and Society*, 2015. https://doi.org/10.1155/2015/657352

Lina, L.R. and Ullah, H. (2019). The concept and implementation of Kaizen in an organization. *Global Journal of Management and Business Research*, 19(A1), 9–17. Retrieved from https://journalofbusiness.org/index.php/GJMBR/article/view/2678

Liu, C., Chai, K.K., Zhang, X., Lau, E.T. and Chen, Y. (2018). Adaptive blockchain-based electric vehicle participation scheme in smart grid platform. *IEEE Access*. doi: 10.1109/ACCESS.2018.2835309

Liu, F., Fan, H.Y. and Qi, J.Y. (2022). Blockchain technology, cryptocurrency: Entropy-based perspective. *Entropy*, 24(4), 557. MDPI AG. https://doi.org/10.3390/e24040557

Liu, J., Zhang, H. and Zhen, L. (2021). *Blockchain Technology in Maritime Supply Chains: Applications, Architecture and Challenges*. Informa UK Limited. https://doi.org/10.1080/00207543.2021.1930239

Loss, T. (2020). *Art Auctions Embrace a Future of Socially Distant Bidding* (Published 2020). https://www.nytimes.com/2020/10/22/arts/auctions-technology.html.

Luecke, M. and Schneiderheinze, C. (2017). More financial burden-sharing for developing countries that host refugees. *Journal of Economics*, 11(1).

Macedo, L. (2018). Blockchain for trade facilitation: Ethereum, eWTP, C.Os, and regulatory issues. *World Customs Journal*, 12(2), 87–94. https://worldcustomsjournal.org/Archives

Mageto, J. and Luke, R. (2020). Skills frameworks: A focus on supply chains. *Journal of Transport and Supply Chain Management*, 14(1), 1–17. http://dx.doi.org/10.4102/jtscm.v14i0.458

Maguire, M. (2000). Policing by risks and targets: Some dimensions and implications of intelligence-led crime control. *Policing and Society*, 9(1), 315–336.

Maguire, M. (2003). Criminal investigation and crime control. *Handbook of Policing*, 2, 430–464.

Maguire, M. (2004). The crime reduction programme: Reflections on the vision and the reality. *Criminal Justice*, 4(3), 213–238.

Maguire, M. and John, T. (2006). Intelligence led policing, managerialism and community engagement: Competing priorities and the role of the national intelligence model in the UK. *Policing and Society*, 16(1), 67–85. Informa UK Limited. https://doi.org/10.1080/10439460500399791

Maheswaran, R.T. and Başar, T. (2003). Nash equilibrium and decentralized negotiation in auctioning divisible resources. *Group Decision and Negotiation*, 12, 361–395. https://doi.org/10.1023/B:GRUP.0000003745.98183.8d

Makani, S., Pittala, R., Alsayed, E., Aloqaily, M. and Jararweh, Y. (2022). A survey of blockchain applications in sustainable and smart cities. *Cluster Computing*, 25(6),

3915–3936. Springer Science and Business Media LLC. https://doi.org/10.1007/s10586-022-03625-z

Man Lai, C., Ka Yin, C., Michael Huen Sum, L., Tse, G., Ka Yan, H., Flint, S.W., Broom, D.R., Ejoe Kar Ho, T. and Ka Yiu, L. (2019). Examining consumers' adoption of wearable healthcare technology: The role of health attributes. *International Journal of Environmental Research and Public Health*, 16(13). https://dx.doi.org/10.3390/ijerph16132257

Marr, B. (2022, November 11). The top 5 technology challenges in 2023. *Forbes*. https://www.forbes.com/sites/bernardmarr/2022/11/10/the-top-5-technology-challenges-in-2023/?sh=86cbc4930c95

McIlgorm, A., Raubenheimer, K., McIlgorm, D.E. and Nichols, R. (2022). The cost of marine litter damage to the global marine economy: Insights from the Asia-Pacific into prevention and the cost of inaction. *Mar. Pollut. Bull.*, 174, 113167 https://doi.org/ 10.1016/j.marpolbul.2021.113167

Mello, J.E., Manuj, I. and Flint, D.J. (2021). Leveraging grounded theory in supply chain research: A researcher and reviewer guide. *International Journal of Physical Distribution & Logistics Management*, 51(10), 1108–1129. https://doi.org/10.1108/IJPDLM-12-2020-0439

Mengelkamp, E., Gärttner, J., Rock, K., Kessler, S., Orsini, L. and Weinhardt, C. (2018). Designing microgrid energy markets: A case study. *The Brooklyn Microgrid, Applied Energy*. https://doi.org/10.1016/j.apenergy.2017.06.054

Mikulic, M. (2020, December 10). *U.S. Economic Loss due to Counterfeit Drugs by Scenario 2020*. Retrieved March 5, 2023, from https://www.statista.com/statistics/1181283/us-cost-due-to-counterfeit-drugs-by-scenario/#:~:text=Based%20on%20estimates%20saying%20the,revenues%20in%20the%20United%20States

Miličević, K., Omrčen, L., Kohler, M. and Lukić, I. (2022). Trust model concept for IoT blockchain applications as part of the digital transformation of metrology. *Sensors*, 22(13), 4708. MDPI AG. https://doi.org/10.3390/s22134708

Milton-Edwards, B. (2003). Iraq, past, present, and future: A thoroughly modern mandate? History and Policy. https://www.historyandpolicy.org/policy-papers/papers/iraq-past-present-and-future-a-thoroughly-modern-mandate

Min, S., Zacharia, Z.G. and Smith, C.D. (2019). Defining supply chain management: In the past, present, and future. *Journal of Business Logistics*, 40(1), 44–55. https://doi.org/10.1111/jbl.12201

Mohit, M., Kaur, S. and Singh, M. (2021). Design and implementation of transaction privacy by virtue of ownership and traceability in blockchain based supply chain. *Cluster Computing*, 25(3), 2223–2240. Springer Science and Business Media LLC. https://doi.org/10.1007/s10586-021-03425-x

Mondal, S., Khatoon, A. and Poria, S. (2020). Inventory modeling and inventory control application. *Supply Chain Intelligence*, 131–154. Springer. https://doi.org/10.1007/978-3-030-46425-7_7

Morrissey, D. (2020, August 18). Security concerns and solutions regarding blockchain use in healthcare. *ATT*. Retrieved March 5, 2023, from https://cybersecurity.att.com/blogs/security-essentials/security-concerns-and-solutions-regarding-blockchain-use-in-healthcare

Mulhim, H. (2022, September 21). Why the crypto industry needs to be AML and KYC compliant. *World Economic Forum*. https://www.weforum.org/agenda/2022/09/whycrypto-businesses-need-anti-money-laundering-regulations/

Mundhe, R. (2022). A study on opportunities and challenges of cryptocurrency. *International Journal of Research in Management & Social Science*, 10(3), 97–99. https://www.researchgate.net/profile/Muthmainnah

Murphy, J., Mermelstein, J., Kokal, T., Tomer, B.K., Hickson, L. Jr. and Sultan, M.M. (2023). Release Number 8647–23. CFTC Charges Avraham Eisenberg with manipulative and deceptive scheme to misappropriate over $110 million from Mango Markets, a digital asset exchange. https://www.cftc.gov/PressRoom/PressReleases/8647-23

Nacci, D., Noferi, M.D. and Santambrogio, D.S. (2015). Danger-system: Exploring new ways to manage occupants safety in intelligent building. *2015 IEEE 2nd World Forum on IoT* (WF-IoT), pp. 675–680.

Narayanan, A. and Ishfaq, R. (2022). Impact of metric-alignment on supply chain performance: A behavioral study. *The International Journal of Logistics Management*, 33(1), 365–384. https://doi.org/10.1108/IJLM-01-2021-0061

National Registry of Exonerations (2023). https://www.law.umich.edu/special/exoneration/Pages/about.aspx

Ngai, E.W., Chau, D.C. and Chan, T.L.A. (2011). Information technology, operational, and management competencies for supply chain agility: Findings from case studies. *The Journal of Strategic Information Systems*, 20(3), 232–249. https://doi.org/10.1016/j.jsis.2010.11.002

Nguyen, M.A.T., Lei, H., Vu, K.D. and Le, P.B. (2019). The role of cognitive proximity on supply chain collaboration for radical and incremental innovation: A study of a transition economy. *Journal of Business & Industrial Marketing*, 34(3), 591–604. https://doi.org/10.1108/JBIM-07-2017-0163

NTT DATA (2022). *Shippers and 3 PLs Work to Rebuild Supply Chains*. Industry/links/57dce54208aeea195935db1c/How-to-Improve-Supply-Chain-Learningin-the-3PL-Industry.pdf

Odeh, A., Keshta, I. and Al-Haija, Q.A. (2022). Analysis of blockchain in the healthcare sector: Application and issues. *Symmetry*, 14(9), 1760. MDPI AG. https://doi.org/10.3390/sym14091760

Ojha, D., Acharya, C. and Cooper, D. (2018a). Transformational leadership and supply chain ambidexterity: Mediating role of supply chain organizational learning and moderating role of uncertainty. *International Journal of Production Economics*, 197, 215–231. https://doi.org/10.1016/j.ijpe.2018.01.001

Ojha, D., Struckell, E., Acharya, C. and Patel, P.C. (2018b). Supply chain organizational learning, exploration, exploitation, and firm performance: A creation-dispersion perspective. *International Journal of Production Economics*, 204, 70–82. https://doi.org/10.1016/j.ijpe.2018.07.025

O'Keeffe, M., Nickel, B., Dakin, T., Maher, C.G., Albarqouni, L., McCaffery, K., Barratt, A. and Moynihan, R. (2021). Journalists' views on media coverage of medical tests and overdiagnosis: A qualitative study. *BMJ Open*, 11(6). https://dx.doi.org/10.1136/bmjopen- 2020-043991

Overstreet, R.E., Skipper, J.B., Huscroft, J.R., Cherry, M.J. and Cooper, A.L. (2019). Multi-study analysis of learning culture, human capital and operational performance in supply chain management. *Journal of Defense Analytics and Logistics*, 3(1), 41–59. https://doi.org/10.1108/JDAL-11-2018-0017

Palizban, O., Kauhaniemi, K. and Guerrero, J.M. (2014). Microgrids in active network management—Part I: Hierarchical control, energy storage, virtual power plants, and market participation. *Renewable and Sustainable Energy Reviews*. https://doi.org/10.1016/j.rser.2014.01.016

Panina, E., Simbuletova, R. and Kakhuzheva, Z. (2022). Analysis of the applicability of blockchain technology in tourism. *SHS Web of Conferences*. https://doi.org/10.1051/shsconf/202214101007

Patel, H.J. (2017). Behavioral aspects of supply chain management: Strategy, commitment, integration and firm performance – A conceptual framework. *International Journal of Supply and Operations Management*, 4(4), 370–375.

Patsonakis, C., Terzi, S., Moschos, I., Ioannidis, D., Votis, K. and Tzovaras, D. (2019). Permissioned blockchains and virtual nodes for reinforcing trust between aggregators and prosumers in energy demand response scenarios. Proceedings of the 2019 IEEE International Conference on Environment and Electrical Engineering and 2019 IEEE Industrial and Commercial Power Systems Europe (EEEIC/I& CPS Europe), Genoa, Italy, 2019, pp. 1–6, https://doi.org/10.1109/EEEIC.2019.8783521

Paul, K.A. (2018). Ancient artifacts vs. digital artifacts: New tools for unmasking the sale of illicit antiquities on the dark web. *Arts*, 7(2), 12. MDPI. https://doi.org/10.3390/arts7020012

Perera, H.N., Fahimnia, B. and Tokar, T. (2020). Inventory and ordering decisions: A systematic review on research driven through behavioral experiments. *International Journal of Operations & Production Management*, 40(7/8), 997–1039. https://doi.org/10.1108/IJOPM-05-2019-0339

Pieroni, C. (2018). La Crypto Nostra: How organized crime thrives in the era of cryptocurrency. *North Carolina Journal of Law & Technology*, 20(5), 111. https://scholarship.law.unc.edu/cgi/viewcontent.cgi?article=1380&context=ncjolt

Porter, M.E. (1979, March/April). How competitive forces shape strategy. *Harvard Business Review*, 137–145.

Posey, C., Raja, U., Crossler, R.E. and Burns, A.J. (2017). Taking stock of organisations' protection of privacy: Categorising and assessing threats to personally identifiable information in the USA. *European Journal of Information Systems*, 26(6), 585–604. https://dx.doi.org/10.1057/s41303-017-0065-y

Prasad, E. (2022, February 14). Digital currencies carry threats as well as promises. *Financial Times*. https://www.ft.com/content/0ba1da77-cbec-48a3-84d2-9682a5d2f1c9

Ratta, P., Kaur, A., Sharma, S., Shabaz, M. and Dhiman, G. (2021). Application of Blockchain and Internet of Things in healthcare and medical sector: Applications, challenges, and future perspectives. *In:* R. Khan (Ed.), *Journal of Food Quality*, 2021, 1–20. Hindawi Limited. https://doi.org/10.1155/2021/7608296

Rathor, S.K. and Saxena, D. (2020) Energy management system for smart grid: An overview and key issues. *International Journal of Energy Research*, 44(6) 4067–4109. https://doi.org/10.1002/er.4883.

Rahmani, M.K.I., Shuaib, M., Alam, S., Siddiqui, S.T., Ahmad, S., Bhatia, S. and Mashat, A. (2022). Blockchain-based trust management framework for cloud computing-based Internet of Medical Things (IoMT): A systematic review. *In:* R. Khan (Ed.), *Computational Intelligence and Neuroscience*, 2022, 1–14. Hindawi Limited. https://doi.org/10.1155/2022/9766844

Razmak, J. and Bélanger, C. (2018). Using the technology acceptance model to predict patient attitude toward personal health records in regional communities. *Information Technology & People*, 31(2), 306–326. https://dx.doi.org/10.1108/ITP-07-2016-0160

Rebolledo, C. and Nollet, J. (2011). Learning from suppliers in the aerospace industry. *International Journal of Production Economics*, 129(2), 328–337. https://doi.org/10.1016/j.ijpe.2010.11.008

Rizzo, N., Sprissler, E., Hong, Y. and Goel, S. (2015). Privacy preserving driving style recognition. *International Conference on Connected Vehicles and Expo 2015*. https://doi.org/10.48550/arXiv.1511.00329

Rodrigues, S.D. and Garcia, V.J. (2023). Transactive energy in microgrid communities: A systematic review. *Science Direct*. https://doi.org/10.1016/j.rser.2022.

Rogers, E.M. (2010). *Diffusion of Innovations*. Simon and Schuster.

Rymarczyk, J. (2022). Transformation of transnational corporations' supply chains as a result of the fourth industrial revolution and the COVID-19 pandemic. *Sustainability*, 14(9), 5518.

Sai, B.D.S., Nikhil, R., Prasad, S. and Naik, N.S. (2023). A decentralised KYC based approach for microfinance using blockchain technology. *Cyber Security and Applications*, 1, 100009. Elsevier BV. https://doi.org/10.1016/j.csa.2022.100009

Sadon, T. (2021). Money laundering: The key to cryptocurrency crime. *Cognyt*. https://www.cognyte.com/blog/anti-money-laundering-cryptocurrency/

Saldamli, G., Upadhyay, C., Jadhav, D., Shrishrimal, R., Patil, B. and Tawalbeh, L. (2022). Improved gossip protocol for blockchain applications. *Cluster Computing*, 25(3), 1915–1926). Springer Science and Business Media LLC. https://doi.org/10.1007/s10586-021-03504-z

Schniederjans, D.G., Curado, C. and Khalajhedayati, M. (2020). Supply chain digitization trends: An integration of knowledge management. *International Journal of Production Economics*, 220, 107439. https://doi.org/10.1016/j.ijpe.2019.07.012

Schorsch, T., Wallenburg, C.M. and Wieland, A. (2017). The human factor in SCM: Introducing meta-theory of behavioral supply chain management. *International Journal of Physical Distribution & Logistics Management*, 47(4), 238–262. https://doi.org/10.1108/IJPDLM-10-2015-0268

Seh, A.H., Zarour, M., Alenezi, M., Sarkar, A.K., Agrawal, A., Kumar, R. and Khan, A.R. (2020). Healthcare data breaches: Insights and implications. *Healthcare*, 8(2), 133. https://doi.org/10.3390/healthcare8020133

Selviaridis, K. and Spring, M. (2018). Supply chain alignment as process: Contracting, learning and pay-for-performance. *International Journal of Operations & Production Management*, 38(3), 732–755. https://doi.org/10.1108/IJOPM-01-2017-0059

Sharma, R. (2020). *Brooklyn Microgrid Gets Approval for Blockchain-based Energy Trading*. https://www.districtenergy.org/blogs/microgrid-resources-coalition/2020/01/06/brooklyn-microgrid-gets-approval-for-blockchain-ba

Sherchan W., Nepal, S. and Paris, C. (2013). A survey of trust in social networks. *ACM Computing Surveys (CSUR)*, 45.4 (2013): 1–33.

Shieber, J. (2017). *Blockchain Consortium R3 Raises $107 million*. https://techcrunch.com/2017/05/23/blockchain-consortium-r3-raises-107-million/

Shou, Y., Hu, W. and Xu, Y. (2018). Exploring the role of intellectual capital in supply chain intelligence integration. *Industrial Management & Data Systems*, 118(5), 1018–1032. https://doi.org/10.1108/IMDS-06-2017-0285

Siano, P., De Marco, G., Rolan, A. and Loia, V. (2019). A survey and evaluation of the potentials of distributed ledger technology for peer-to-peer transactive energy exchanges in local energy markets. *IEEE Systems Journal*, 13(3), 3454–3466. https://doi.org/10.1109/JSYST.2019.2903172

Singh, P., Jain, R.K. and Sharma, R. (2021). Application of social psychology in supply chain management: A review. *International Journal of Management, Economics and Social Sciences*, 10(2–3), 110–127. http://dx.doi.org/10.32327/IJMESS/10.2-3.2021.7

Sjöström, J., Ågerfalk, P. and Hevner, A.R. (2022). The design of a system for online psychosocial care: Balancing privacy and accountability in sensitive online healthcare environments. *Journal of the Association for Information Systems*, 23(1), 237–263. https://doi.org/10.17705/1jais.00717

Solodan, K. (2019). Legal regulation of cryptocurrency taxation in European countries. *Eur. J.L. & Pub. Admin.*, 6, 64. https://doi.org/10.18662/eljpa/64

Southworth Davis, K.A., Bashford, O., Jewell, A., Shetty, H., Stewart, R.J., Sudlow, C.L.M. and Hotopf, M.H. (2018). Using data linkage to electronic patient records to assess the validity of selected mental health diagnoses in English Hospital Episode Statistics (HES). *PLoS One*, 13(3). https://dx.doi.org/10.1371/journal.pone.0195002

Stabenow, D. (2022). S.4760 – 117th Congress (2021-2022): Digital Commodities Consumer Protection Act of 2022 (2022, September 15). https://www.congress.gov/bill/117th-congress/senate-bill/4760/cosponsors

Stanley, A. (2018, August 27). PwC: Regulatory uncertainty is largest impediment to blockchain adoption. *Forbes*. https://www.forbes.com/sites/astanley/2018/08/27/pwc-regulatory-uncertainty-is-largest-impediment-to-blockchain-adoption/?sh=131a3650b014

Sterman, J.D. (1989). Modeling managerial behavior: Misperceptions of feedback in a dynamic decision making experiment. *Management Science*, 35(3), 321–339. https://doi.org/10.1287/mnsc.35.3.321

Sterman, J., Oliva, R., Linderman, K.W. and Bendoly, E. (2015). System dynamics perspectives and modeling opportunities for research in operations management. *Journal of Operations Management*, 39(40), 1–5. https://doi.org/10.1016/j.jom.2015.07.001

Strauss, L.J. (2018). Enforcing the privacy and security rules. *Journal of Health Care Compliance*, 20(4), 55–58.

Sun, J., Yan, J. and Zhang, K.Z.K. (2016). Blockchain-based sharing services: What blockchain technology can contribute to smart cities. *Financial Innovation*, 2(1), 1–9.

Susilowardhani, S., Bidari, A.S. and Nurviana, R. (2022). Regulation and the future of cryptocurrency in Indonesia. *International Journal of Economics, Business, and Accounting Research (IJEBAR)*, 6(3), 1568–1572.

Sutia, S., Riadi, R. and Fahlevi, M. (2020). The influence of supply chain performance and motivation on employee performance. *International Journal of Supply Chain Management*, 9(2), 86–92.

Sweeney, E., Evangelista, P. and Passaro, R. (2005). Putting supply-chain learning theory into practice: Lessons from an Irish case. *International Journal of Knowledge and Learning*, 1(4), 357–372. https://www.researchgate.net/profile/Renato-

Tan, Z., Wu, N., Zheng, Y. and Chen, W. (2022). Study on the influence of grid connected capacity difference of photovoltaic power generation on power grid performance. *Journal of Physics*. https://doi.org/10.1088/1742-6596/2310/1/012003

Tang, W., Ren, J. and Zhang, Y. (2019). Enabling trusted and privacy-preserving healthcare services in social media health networks. *IEEE Transactions on Multimedia*, 21(3), 579–590. https://doi.org/10.1109/TMM.2018.2889934

Tanwar, S., Batra, S., Gupta, M. and Rana, A. (2021). Efficient and secure multiple digital signatures to prevent forgery based on ECC. *International Journal of Applied Science and Engineering*, 18(5), 1–7. https://doi.org/10.6703/ijase.202109_18(5).010

Tatham, P., Wu, Y., Kovács, G. and Butcher, T. (2017). Supply chain management skills to sense and seize opportunities. *The International Journal of Logistics Management*, 28(2), 266–289. https://doi.org/10.1108/IJLM-04-2014-0066

Teniwut, W. and Hasyim, C. (2020). Decision support system in supply chain: A systematic literature review. *Uncertain Supply Chain Management*, 8(1), 131–148. https://doi.org/10.5267/j.uscm.2019.7.009

Thakkar, J., Kanda, A. and Deshmukh, S.G. (2011). Mapping of supply chain learning: A framework for SMEs. *The Learning Organization*, 18(4), 313–332. https://doi.org/10.1108/09696471111132522

Tierion: Blockchain Proof Engine (2023). Tierion.com. https://tierion.com/

Tokar, T., Aloysius, J.A. and Waller, M.A. (2012). Supply chain inventory replenishment: The debiasing effect of declarative knowledge. *Decision Sciences*, 43(3), 525–546. https://doi.org/10.1111/j.1540-5915.2012.00355.x

Toradmalle, D., Muthukuru, J. and Sathyanarayana, B. (2019). Certificateless and a provably secure digital signature scheme based on an elliptic curve. *International Journal of Electrical and Computer Engineering (ICE)*, 9(4), 3228.

Toscher, S. and Stein, M.R. (2018). Cryptocurrency-FinCEN and discovery of hidden wealth. *J. Tax Prac. & Proc.*, 20, 19. https://www.taxlitigator.com/wpcontent/uploads/2018/10/Hidden_Wealth.pdf

Tsai, F.C. (2021). The application of Blockchain of custody in criminal investigation process. *Procedia Computer Science*, 192(1), 2779–2788.

Turner, A. and Irwin, A.S.M. (2018, January 2). Bitcoin transactions: A digital discovery of illicit activity on the blockchain. *Journal of Financial Crime*, 25(1), 109–130. https://doi.org/10.1108/jfc-12-2016-0078

Underwood, S. (2016). Blockchain beyond bitcoin. *Communications of the ACM*, 59(11), 15–17.

Vailshery, L.S. (2022). *Number of Internet of Things (IoT) Connected Devices Worldwide 2019–2021*. https://www.statista.com/statistics/1183457/iot-connected-devices-worldwide/

Van Cutsem, O., Dac, D.H., Boudou, P. and Kayal, M. (2019). Cooperative energy management of a community of smart-buildings: A blockchain approach. *International Journal of Electrical Power and Energy Systems*. https://doi.org/10.1016/j.ijepes.2019.105643.

van Hoek, R. (2021). Lessons from CSCMP supply chain hall of famer Henry Ford and the research that they call for in modern supply chains. *International Journal of Physical Distribution & Logistics Management*, (52)1, 88–102. https://doi.org/10.1108/IJPDLM- 10-2020-0315

Venkatesh, V.G., Kang, K., Wang, B., Zhong, R.Y. and Zhang, A. (2020). *System Architecture for Blockchain based Transparency of Supply Chain Social Sustainability*. Elsevier BV. https://doi.org/10.1016/j.rcim.2019.101896

Vervoort, D., Tam, D.Y. and Wijeysundera, H.C. (2022). Health technology assessment for cardiovascular digital health technologies and artificial intelligence: Why is it different? *Can. J. Cardiol.*, 38(2), 259–266. https://doi.org/10.1016/j.cjca.2021.08.015

Viet, N.Q., Behdani, B. and Bloemhof, J. (2018). The value of information in supply chain decisions: A review of the literature and research agenda. *Computers & Industrial Engineering*, 120, 68–82. https://doi.org/10.1016/j.cie.2018.04.034

Viswanath, S.K., Yuen, C., Tushar, W., Li, W.T., Wen, C.K., Hu, K., Chen, C. and Liu, X. (2016). System design of the IoT for residential intelligent grid. *IEEE Wirel. Commun.*, 23(5), 90–98.

Wahab, S.N., Bahar, N. and Radzi, N.A.M. (2021). An inquiry on knowledge management in third-party logistics companies. *International Journal of Business Innovation and Research*, 24(1), 124–146. https://doi.org/10.1504/IJBIR.2021.111977

Walport, M. (2016). *Distributed Ledger Technology: Beyond Blockchain.* https://assets. publishing.service.gov.uk/government/uploads/system/uploads/attachment_data/ file/492972/

Wamba, S.F., Queiroz, M.M. and Trinchera, L. (2020). Dynamics between blockchain adoption determinants and supply chain performance: An empirical investigation. *Int. J. Prod. Econ.*, 229, 107791.

Wang, C., Hao, S. and Ma, Y. (2022). Blockchain-based intelligent interconnection system optimization decision. *In:* J. Liu (Ed.), *Security and Communication Networks*, 2022, 1–12. Hindawi Limited. https://doi.org/10.1155/2022/6818562

Wang, J., Liu, J., Wang, F. and Yue, X. (2021). *Blockchain Technology for Port Logistics Capability: Exclusive or Sharing.* Elsevier BV. https://doi.org/10.1016/j. trb.2021.05.010

Wang, L. and Liu, F. (2022). Integration and dissemination of sports big data based on blockchain. *Mobile Information Systems.* https://doi.org/10.1155/2022/3208904

Wang, S., Zhang, S. and Zhang, Y. (2018). A blockchain-based framework for data sharing with fine-grained access control in decentralized storage systems. *IEEE Access*, 6, 38437–38450.

Wang, S. and Zhu, X. (2021). Evaluation of potential cryptocurrency development ability in terrorist financing. *Policing: A Journal of Policy and Practice*, 15(4), 2329–2340. https://doi.org/10.1093/police/paab059

Wang, Y., Singgih, M., Wang, J. and Rit, M. (2019). Making sense of blockchain technology: How will it transform supply chains? *Int. J. Prod. Econ.*, 211, 221–236.

What Does Proof-of-Stake (PoS) Mean in Crypto? (2023). *Investopedia.* https://www. investopedia.com/terms/p/proof-stake-pos.asp

What Is Proof of Work (PoW) in Blockchain? (2023). *Investopedia.* https://www. investopedia.com/terms/p/proof-work.asp#:~:text=Key%20Takeaways,a%20 reward%20for%20work%20done.

Wiewiora, A., Smidt, M. and Chang, A. (2019). The 'how' of multi-level learning dynamics: A systematic literature review exploring how mechanisms bridge learning between individuals, teams/projects and the organization. *European Management Review*, 16(1), 93–115. https://doi.org/10.1111/emre.12179

Wong, D.R., Bhattacharya, S. and Butte, A.J. (2019). Prototype of running clinical trials in an untrustworthy environment using blockchain. *Nature Communications*, 10(1), 1–8.

Wong, L.W., Tan, G.W.H., Lee, V.H., Ooi, K.B. and Sohal, A. (2020). Unearthing the determinants of Blockchain adoption in supply chain management. *Int. J. Prod. Res.*, 58(7), 2100–2123.

Wu, M. and Liu, H. (2020). Integration of IoT and blockchain for movable property asset pledge financial service. 19th International Symposium on *Distributed Computing and Applications for Business Engineering and Science (DCABES).* DOI: 10.1109/ DCABES50732.2020.00051

Wyman, R.E. (2020). Can HIPAA be saved? The continuing relevance and evolution of healthcare privacy and security standards. *The Health Lawyer*, 32(6), 5–23.

Xiao, Y., Zhang, H., Yuan, C., Gao, N., Meng, Z. and Peng, K. (2020). The design of an intelligent high-speed loom industry interconnection remote monitoring system. *Wireless Personal Communications*, 1–21.

Xie, Z., Kong, H. and Wang, B. (2022). Dual-chain Blockchain in agricultural e-commerce information traceability considering the Viniar algorithm. *In:* L. Sun (Ed.), *Scientific Programming*, 2022, 1–10. Hindawi Limited. https://doi.org/10.1155/2022/2604216

Xu, L., Di, Z. and Chen, J. (2021). *Evolutionary Game of Inland Shipping Pollution Control Under Government Co-Supervision.* Elsevier BV. https://doi.org/10.1016/j.marpolbul.2021.112730

Xue, X. (2022). Design of enterprise financial information fusion sharing system based on Blockchain Technology. *In:* D. Plewczynski (Ed.), *Computational Intelligence and Neuroscience*, 2022, 1–12. Hindawi Limited. https://doi.org/10.1155/2022/5402444

Xuegang, B., Peng, H. and Zhanbo S. (2011). Real time queue length estimation for signalized intersections using travel times from mobile sensors. *Transportation Research Part C: Emerging Technologies*, 19(6), 1133–1156, ISSN 0968-090X, https://doi.org/10.1016/j.trc.2011.01.002

Yaga, D., Mell, P., Roby, N. and Scarfone, K. (2019). *Blockchain Technology Overview.* arXiv preprint arXiv:1906.11078.

Yang, A., Gao, X., Zhang, H. and Massa, M. (2019). R3: Putting the 'Fin' back in FinTech. https://hbsp.harvard.edu/product/IN1544-PDF-ENG?activeTab=overview

Yang, J., Wang, Z. and Zhang, X. (2015). An ibeacon-based indoor positioning systems for hospitals. *Int. J. Intelligent Home*, 9(7), 161–168.

Yang, Y., Jia, F. and Xu, Z. (2019). Towards an integrated conceptual model of supply chain learning: An extended resource-based view. *Supply Chain Management: An International Journal*, 24(2), 189–214. https://doi.org/10.1108/SCM-11-2017-0359

Yavaprabhas, K., Pournader, M. and Seuring, S. (2022). Blockchain as the "trust-building machine" for supply chain management. *Annals of Operations Research.* Springer Science and Business Media LLC. https://doi.org/10.1007/s10479-022-04868-0

Yiannas, F. (2017). A new era of food transparency with Wal-Mart center in China. *Food Saf. News.* Retrieved in April, 2021 from https://www.foodsafetynews.com/2017/03/a-new-era-of-food-transparency-with-wal-mart-center-in-china/#:~:text=Wal%2DMart%20and%20the%20Wal,its%20plans%20for%20the%20year.

Ying, L. and Gui-hang, G. (2021). Overconfident newsvendor's joint pricing and ordering decisions under the additive demand case. *International Journal of Business and Management*, 13(11), 135–135. https://doi.org/10.5539/ijbm.v13n11p135

Your Health At Your Fingertips (2023). Patientory.com. https://patientory.com/about

Zainudin, Z. (2021). More targeted cryptocurrency attacks, advanced threats in 2022 – Kaspersky. Bernama: *Malaysian National News Agency.*

Zakari, N., Al-Razgan, M., Alsaadi, A., Alshareef, H., Al Saigh, H., Alashaikh, L., Alharbi, M. and Alotaibi, S. (2022). Blockchain technology in the pharmaceutical industry: A systematic review. *Peer J. Computer Science*, 8, e840.

Zarowitz, B.J. (2020). Emerging pharmacotherapy and health care needs of patients in the age of artificial intelligence and digitalization. *Ann. Pharmacother.*, 54(10), 1038–1046. https://doi.org/10.1177/1060028020919383

Zehir, C., Zehir, M., Borodin, A., Mamedov, Z.F. and Qurbanov, S. (2022). Tailored blockchain applications for the natural gas industry: The case study of SOCAR. *Energies*, 15(16), 6010. MDPI AG. https://doi.org/10.3390/en15166010

Zhang, J., Tan, R., Su, C. and Si, W. (2020). Design and application of a personal credit information sharing platform based on consortium blockchain. *Journal of Information Security and Applications*, 55, 102659.

Zhang, W., Yang, W., Chen, C., Li, N. and Bao, Z. (2022). Toward privacy-preserving blockchain-based electricity auction for V2G networks in the smart grid. *Security and Communications Networks.* https://doi.org/10.1155/2022/6911463

Zhao, H., Liu, J., Zhao, P. and Chen, J. (2022). Will nuclear polluted seafood stop selling in the blockchain-enabled market? Lessons from government punishment and social cognition for retailer's selling. *Marine Pollution Bulletin*, 178, 113608. Elsevier BV. https://doi.org/10.1016/j.marpolbul.2022.113608

Zheng, Z., Xie, S., Dai, H., Chen, X. and Wang, H. (2017, June). An overview of blockchain technology: Architecture, consensus, and future trends. *2017 IEEE International Congress on Big Data (BigData Congress)*, 557–564. IEEE.

Zhijian, Z., Wang, P., Wan, M., Guo, J. and Liu, J. (2020). Supply chain decisions and coordination under the combined effect of overconfidence and fairness concern. *Complexity*, 2020, 1–16. https://doi.org/10.1155/2020/3056305

Zhu, Q., Krikke, H. and Caniels, M.C. (2018). Supply chain integration: Value creation through managing inter-organizational learning. *International Journal of Operations & Production Management*, 38(1), 211–229. https://doi.org/10.1108/IJOPM-06-2015-0372

Zhu, Y. (2022). Research on real-time tracking algorithm of e-commerce logistics information based on Blockchain Technology. *In:* J. Ye (Ed.), *Computational Intelligence and Neuroscience*, 2022, 1–13. Hindawi Limited. https://doi.org/10.1155/2022/7006506

Index

For Product Safety Concerns and Information please contact our EU
representative GPSR@taylorandfrancis.com
Taylor & Francis Verlag GmbH, Kaufingerstraße 24, 80331 München, Germany

9 7 8 1 0 3 2 4 7 1 5 7 0